Juha Taina

Databases in Mobile Telecommunications

Juha Taina

Databases in Mobile Telecommunications

Architectures and Performance Analysis

VDM Verlag Dr. Müller

Impressum/Imprint (nur für Deutschland/ only for Germany)
Bibliografische Information der Deutschen Nationalbibliothek: Die Deutsche Nationalbibliothek
verzeichnet diese Publikation in der Deutschen Nationalbibliografie; detaillierte bibliografische
Daten sind im Internet über http://dnb.d-nb.de abrufbar.
Alle in diesem Buch genannten Marken und Produktnamen unterliegen warenzeichen-, marken-
oder patentrechtlichem Schutz bzw. sind Warenzeichen oder eingetragene Warenzeichen der
jeweiligen Inhaber. Die Wiedergabe von Marken, Produktnamen, Gebrauchsnamen,
Handelsnamen, Warenbezeichnungen u.s.w. in diesem Werk berechtigt auch ohne besondere
Kennzeichnung nicht zu der Annahme, dass solche Namen im Sinne der Warenzeichen- und
Markenschutzgesetzgebung als frei zu betrachten wären und daher von jedermann benutzt
werden dürften.

Coverbild: www.purestockx.com

Verlag: VDM Verlag Dr. Müller Aktiengesellschaft & Co. KG
Dudweiler Landstr. 125 a, 66123 Saarbrücken, Deutschland
Telefon +49 681 9100-698, Telefax +49 681 9100-988, Email: info@vdm-verlag.de
Zugl.: Helsinki, University of Helsinki, 2003

Herstellung in Deutschland:
Schaltungsdienst Lange o.H.G., Zehrensdorfer Str. 11, D-12277 Berlin
Books on Demand GmbH, Gutenbergring 53, D-22848 Norderstedt
Reha GmbH, Dudweiler Landstr. 99, D- 66123 Saarbrücken
ISBN: 978-3-639-09284-4

Imprint (only for USA, GB)
Bibliographic information published by the Deutsche Nationalbibliothek: The Deutsche
Nationalbibliothek lists this publication in the Deutsche Nationalbibliografie; detailed
bibliographic data are available in the Internet at http://dnb.d-nb.de.
Any brand names and product names mentioned in this book are subject to trademark, brand or
patent protection and are trademarks or registered trademarks of their respective holders. The use
of brand names, product names, common names, trade names, product descriptions etc. even
without
a particular marking in this works is in no way to be construed to mean that such names may be
regarded as unrestricted in respect of trademark and brand protection legislation and could thus
be used by anyone.

Cover image: www.purestockx.com

Publisher:
VDM Verlag Dr. Müller Aktiengesellschaft & Co. KG
Dudweiler Landstr. 125 a, 66123 Saarbrücken, Germany
Phone +49 681 9100-698, Fax +49 681 9100-988, Email: info@vdm-verlag.de

Produced in USA and UK by:
Lightning Source Inc., 1246 Heil Quaker Blvd., La Vergne, TN 37086, USA
Lightning Source UK Ltd., Chapter House, Pitfield, Kiln Farm, Milton Keynes, MK11 3LW, GB
BookSurge, 7290 B. Investment Drive, North Charleston, SC 29418, USA
ISBN: 978-3-639-09284-4

Contents

Chapter 1

Introduction

Specialization and reusability are the key components in current software engineering. High-quality software must be created efficiently and tailored for very specific application environments. Yet often the requirements and architecture analysis are incomplete or missing. It is indeed difficult to analyze a complex piece of software from incomplete specifications.

One interesting area of specifically tailored software of standard components is arising in database management systems. It is no longer sufficient to create a monolithic database management system and then sell it to everyone. Instead, customers are interested in tailored database solutions that offer services they need. The better the database management systems are tailored for the customers, the smaller and more efficient they may become.

Unfortunately even a very good specification of database management needs may fail if the final product cannot justify data integrity and transaction throughput requirements. Without good analysis tools we may be in a project of creating millions of lines of code for a product that can never be used. Bottlenecks must be identified and solved before we go from requirements analysis to design.

We are especially interested in telecommunications databases, an area which has been considered very interesting both in the telecommunications and computer science research field. When database technology became mature enough to support telecommunications, new requirements soon started to arise. For instance, in his article, Ahn lists typical telecommunications database tasks including such topics as maintaining reference data, processing traffic data, ability to process temporal queries, and high availability [Ahn94]. This information, while specific, is only a wish list from customers to database management system designers.

While telecommunications database research has mostly concentrated on embedded data managers and their requirements, little work has been done on defining and analyzing a stand-alone database architecture for telecommunications. A typical example of an embedded data manager architecture is the IN solution for fixed and cellular networks from Nokia, as first described in 1995 [LW95]. While this is an interesting approach, it does not benefit from the clear distributed environment of the network. Such embedded systems do not form a single distributed database nor are they very fruitful for a queueing model analysis.

The embedded system solution distributes and replicates data to several small embedded databases. An alternative approach is to use a centralized fault-tolerant IN database. The best known research in this area is carried out at the University of Trondheim, Department of Computer and Information Science. Their Database systems research group is interested in creating a very high throughput and fault-tolerant database management system. The starting point of the system is first described in 1991 [BS91]. Lately their research has concentrated on temporal object-oriented databases [NB00].

Projects Darfin and Rodain at the University of Helsinki, Department of Computer Science have produced a lot of interesting research results of real time fault-tolerant object-oriented databases for telecommunications. The most complete summary of Darfin/Rodain work is that of the Rodain database management system architecture [TR96]. The latest results concentrate on distribution, main memory and telecommunications [LNPR00].

In addition to the telecommunication databases, both real-time and distributed databases have been an active research subject for several decades. The general result of combining real time and distribution in databases is clear: trying to add both in a general case does not work. Fortunately for telecommunications databases we do not have to worry about the most difficult real-time and distribution combinations. It is sufficient to see how telecommunications transactions behave in a distributed environment.

In this thesis we define our database management system architecture for Intelligent Networks (IN) and Global System for Mobile Communications (GSM). We define a toolbox of queueing model formulas for transaction based systems and use the formulas for a detailed bottleneck analysis of the IN/GSM database management system. In the analysis we use simple queueing model tools that we have developed for this type of database management system analysis. We show that very large systems can be analyzed efficiently with our toolbox.

The rest of the thesis is structured in five chapters. Chapter 2 gives background information on databases, Intelligent Networks, and current and future mobile networks. In Chapter 3 we present our IN and GSM data analysis. In Chapter 4 we introduce our IN/GSM database management system architecture along with interesting areas of concurrency control and global transaction atomicity issues. In Chapter 5 we first introduce our derived queueing model for transaction-based bottleneck analysis and then show how the model can be efficiently used in analyzing a very complex database management system. The analysis is included, along with the results. Finally, Chapter 6 summarizes our work.

Main contributions

In the following we list the main contributions of the thesis.

1. We define a queueing model toolbox for transaction based systems. The model offers simple yet comprehensive formulas for analyzing various types of systems that use transactions for data management and system execution. The formulas are defined to be useful for analyzing bottlenecks of database management systems.

2. We define our real-time distributed database reference architecture for IN/GSM and give a detailed analysis of the bottlenecks and transaction execution times of the system using our queueing model tools.

3. We give a detailed IN and GSM data and transaction analysis that is based on the IN recommendations and GSM standardization. We use the analysis results in the IN/GSM database analysis.

4. We show in the analysis that distribution is an over-valued aspect in an IN/GSM database. Instead, parallelism can be very useful in the database.

Chapter 2

Background

2.1 General database theory

A *database* is a collection of data that is managed by a software called a database management system (DBMS). The DBMS is responsible for database data maintenance so that database clients can have reliable access to data. The DBMS must also ensure that database data is consistent with its real world equivalence. This property of database data is called *data consistency*.

The only way to access database data is with a *transaction*. A transaction is a collection of database operations that can read and write data in a database. It is similar to a process in an operating system. Like processes in operating systems, transactions have identities and may have a priority. Transactions are the only way to access data in a DBMS.

When a transaction executes in a DBMS, the operations of the transaction are either all accepted or none accepted. If the operations are accepted, the transaction has committed. If they are not accepted, the transaction is aborted. Only after a commit the changes made by the transaction are visible in the database.

A DBMS needs to support certain transaction properties to offer a reasonable access to the database. Usually the properties are listed in four aspects of transactions: atomicity, consistency, isolation and durability. These are referred to as the *ACID-properties* of transactions.

- *Transaction atomicity.* Atomicity refers to the fact that a transaction is a unit of operations that are either all accepted or none accepted. The DBMS is responsible for maintaining atomicity by handling failure situations correctly when transactions execute in parallel.

- *Transaction consistency.* The consistency of a transaction defines its correctness. A correct transaction transfers a database from one consistent state to another. An interesting classification of consistency can be used with the general transaction consistency theory. This classification groups databases into four levels of consistency [GLPT76]. The classification is based on a concept of dirty data. A data item is called dirty if an uncommitted transaction has updated it. Using this definition, the four levels of consistency are as follows.

  - *Degree 4.* Transaction T sees degree 4 consistency if
    1. T does not overwrite dirty data of other transactions,
    2. T does not commit any writes until it completes all its writes,
    3. T does not read dirty data from other transactions, and
    4. other transactions do not dirty any data read by T before T completes.

  - *Degree 3.* Transaction T sees degree 3 consistency if
    1. T does not overwrite dirty data of other transactions,
    2. T does not commit any writes until it completes all its writes, and
    3. T does not read dirty data from other transactions.

  - *Degree 2.* Transaction T sees degree 2 consistency if
    1. T does not overwrite dirty data of other transactions, and
    2. T does not commit any writes until it completes all its writes.

  - *Degree 1.* Transaction T sees degree 1 consistency if
    1. T does not overwrite dirty data of other transactions.

- *Transaction isolation.* Isolation states that each transaction is executed in the database as if it alone had all resources. Hence, a transaction does not see the other concurrently executing transactions.

- *Transaction durability.* Durability refers to transaction commitment. When a transaction commits, the updates made by it to the database are permanent. A new transaction may nullify the updates but only after it has committed. Neither uncommitted transactions nor the system itself can discard the changes. This requirement is important since it ensures that the results of a committed transaction are not lost for some unpredictable reason.

A good DBMS must support ACID-properties. Theoretically the simplest way to support the properties is to let every transaction access the entire database alone. This policy supports atomicity, since a transaction never suffers interference from other transactions, consistency, since the database cannot become inconsistent when transactions cannot access uncommitted data, isolation since each transaction has undisturbed access to the database, and durability since the next transaction is not allowed to execute before the previous transaction has been committed or aborted. However, serial access to the database is not a good approach except to very small databases since it seriously affects resource use. Whenever a transaction is blocked, for instance when disk access is needed, the whole database is unaccessible.

A better alternative than forcing serial access to the database is to allow transactions to execute in parallel while taking care that they maintain data consistency in the database. When the DBMS supports data consistency, transaction operations may be interleaved. This is modeled by a structure called a *history*.

A history indicates the order in which operations of transactions were executed relative to each other. It defines a partial order of operations since some operations may be executed in parallel. If a transaction T_i specifies the order of two of its operations, these two operations must appear in that order in any history that includes T_i. In addition, a history specifies the order of all *conflicting operations* that appear in it.

Two operations are said to *conflict* if the order in which they are executed is relevant. For instance, when we have two read operations, the result of the operations is the same no matter which one is executed first. If we have a write operation and a read operation, it matters whether the read operation is executed before or after the write operation.

A history H is said to be *complete* if for any transaction T_i in H the last operation is either abort or commit. Thus, the history includes only transactions that are completed.

While histories and complete histories are useful tools for analyzing transaction effects on each other, they can also become very complex since they include all transaction operations. Yet usually we are mostly interested in conflicting operations and their effects. Due to this a simpler tool called *serialization graph (SG)* [BHG87] is derived from histories.

A serialization graph *SG* of history H, denoted as $SG(H)$, is a directed graph whose nodes are transactions that are committed in H, and whose edges are all $T_i \rightarrow T_j$ $(i \neq j)$ such that one of T_i's operations precedes and conflicts with one of T_j's operations in H.

In order to display histories we use the following definitions. Let us define o_i to be an operation o for transaction T_i, and $\hat{o}_j$ be an operation $\hat{o}$ for transaction T_j. Let the set of possible operations be {r,w,c,a} where r=read, w=write, c=commit, and a=abort. When operation o_i precedes operation $\hat{o}_j$, operation o_i is listed before operation $\hat{o}_j$.

For example, let us have the following history.

$$H = r_1[x]w_2[x]r_2[y]c_2w_3[y]c_3w_1[x]c_1$$

The history includes transactions T_1, T_2, T_3. It is a complete history the last operation of each transaction is commit. The serialization graph of the history is

$$SG(H) = T_1 \rightleftarrows T_2 \rightarrow T_3.$$

The serialization graphs (and histories as well) are used in serialization theory. A history H is said to be *serializable* if its committed projection, $C(H)$, is equivalent to a serial history H_s [BHG87]. The equivalence here states that the effects of the transactions in $C(H)$ are equal to some serial execution order of the same transactions.

A serializable history causes the same changes and results to a database as some serial execution order of the same transactions. In other words, serializable transactions preserve consistency of the database since serial transactions do the same.

In order to recognize serializable transaction execution orders, we need to examine histories and serialization graphs. A history H is serializable, if and only if its serialization graph $SG(H)$ is acyclic. This is the fundamental theorem of serializability theory. It can be found for instance from the book by Bernstein, Hadzilacos, and Goodman, pp. 33 [BHG87].

For instance, in the previous example, the serialization graph contains a cycle $T_1 \rightarrow T_2 \rightarrow T_1$. Hence, the history is not serializable.

Serializable transactions manage atomicity, consistency, and integrity of ACID-properties. Durability is not necessarily met in serializable transactions which can be seen in the following history:

$$H = w_1[x]r_2[x]c_2a_1$$

History H is complete and its serialization graph $SG(H)$ is acyclic. Yet the history does not preserve durability. Since T_1 is aborted, T_2 should be aborted as well. Unfortunately this is not possible. Transaction T_2 has committed so it no longer exists in the system.

As the example shows, we must take extra care to preserve durability in histories. This aspect of transactions is called *transaction recoverability*.

The recovery system of a DBMS should force the database to contain all effects of committed transactions and no effects of unfinished and aborted transactions. If every transaction will eventually commit, all the DBMS has to do is to allow transactions to execute. No recovery is needed. Hence, the recovery is needed only for managing the effects of aborted transactions.

When a transaction aborts, the DBMS must nullify all changes that the transaction has made to the database. This implies two types of changes: changes to data and changes to other transactions that have read data that has been updated by the aborted transactions.

The DBMS can nullify the changes to data by restoring the old values. The changes to transactions is more complex. If transaction T_1 reads data written by transaction T_2 and T_2 aborts, T_1 must also be aborted. Moreover, if transaction T_3 reads data written by transaction T_1, it must also be aborted even if it does not read data written by T_2. This causes a *cascading abort* where a chain of transactions must be aborted in order to nullify the effects of one aborting transaction. Although a DBMS does not necessarily have to avoid cascading aborts, they are generally not acceptable. The cascading abort chain can be of arbitrary length and every abort wastes resources.

When the DBMS allows a transaction to commit, it at the same time guarantees that all the results of the transactions are relevant. This is a stronger guarantee than one would first think. For instance, let us take the previous example with transactions T_1 and T_2. The initial conditions are the same but now transaction T_2 commits. When transaction T_1 is aborted, the results of transaction T_2 should be nullified. Unfortunately this is not possible, since durability states that the changes T_2 has made to the database are permanent. This execution order is not *recoverable.*

A DBMS can tolerate cascading aborts, but due to the nature of the durability, it cannot tolerate non-recoverable execution orders. Hence, all execution orders of transactions should always be recoverable. Fortunately executions that avoid cascading aborts are a true subset of executions that are recoverable [BHG87]. Thus, if a DBMS supports histories that avoid cascading aborts, it automatically supports recoverable histories.

We can avoid both non-recoverable executions and cascading aborts if we require that the execution of a write operation to a data item x can be delayed until all transactions that have previously written to x are either committed or aborted, and that all read operations to x are delayed until all transactions and did writes to x are either committed and aborted. Together these define a *strict* transaction execution order. A strict order is basically the same than the degree 2 consistency mentioned earlier.

2.2   Real-time databases

In a real-time database management system (RT-DBMS) both data and transactions may have timing constraints. A timing constraint states that an event must happen before or at a specific time. Otherwise the element where the timing constraint is set, is invalidated. In data, a missed timing constraint states that the data item becomes invalid. In transaction, it states that the transaction cannot fulfill its task.

A data item that has a timing constraint is called *temporal*. A RT-DBMS must support the operations and consistency of both static and temporal data. Temporal data is stored information that becomes outdated after a certain period of time [Ram93].

Real-time systems are used in environments where exact timing constraints are needed; databases are used in environments where logical data consistency is needed. Together these requirements state when a real-time database is useful. As listed in [Ram93],

- database schemas help to avoid redundancy of data as well as of its description;

- data management support, such as indexing, assists in efficient access to the data; and

- transaction support, where transactions have ACID-properties, assist in efficient concurrent application use.

Temporal and logical consistency can be divided into two aspects: data consistency and transaction consistency [CDW93]. The former defines data issues. The latter defines data access issues.

Data-temporal consistency deals with the question of *when* data in a database is valid. Static data is always valid. Temporal data is valid if it was previously updated within a predefined time period. Data-logical consistency deals with the question of *how* data is consistent. A data entity is consistent if both its value and references to it are consistent with the rest of the database.

Transaction-temporal consistency deals with the question of *when and how long* a transaction is valid. Transaction-temporal consistency is maintained with deadlines that define when at the latest a transaction should commit. Transaction-logical consistency deals with the question of *how* transactions can execute without interfering with each other for more than allowed. Usually the appropriate interference level is no interference. In such a case, transactions may access data as long as their access histories can be ordered so that they could have been executed alone in the database.

From our analysis' point of view, the most interesting entity in a RT-DBMS is a concurrency controller. Yu, Wu, Lin, and Son have written an excellent survey of concurrency control and scheduling in real-time databases [YWLS94]. In the article they list concurrency controller tasks such as conflict detection, conflict resolution, serialization order and run policy.

The concurrency controller is responsible for maintaining logical consistency at a predefined level. The temporal consistency level is more a design decision than a concurrency controller requirement since temporal consistency constraints are derived from the timing constraints outside the database. Often a RT-DBMS that is logically consistent is also temporally consistent. However, this is not a requirement. Temporally correct serializable schedules are a subset of all serializable schedules [Ram93].

Generally we have two approaches for concurrency control: the pessimistic approach and the optimistic approach. In a pessimistic concurrency control the concurrency controller assumes that a conflict will occur and checks this as early as possible. In an optimistic concurrency control the concurrency controller assumes that no conflicts will occur and checks this assumption as late as possible. Pessimistic methods cause blocking where a transaction must wait for other transactions to release resources. Optimistic methods cause restarts where a transaction must be started over after a data access conflict has been detected.

When an access conflict is detected in a pessimistic concurrency controller, the controller must either block or abort one of the conflicting transactions. In traditional database management systems the latter transaction is usually blocked until the resource is free to use. In a RT-DBMS this policy may lead to priority inversion where a high priority transaction is waiting for a low priority transaction to free resources. The simplest way to resolve a priority inversion conflict is to abort the low priority transaction. This wastes the resources that the lower priority transaction has already used. Sometimes the aborted transaction is very close to the finishing point, in which case it would be better to let the transaction finish its execution.

Next to aborting the low priority transaction, the other way to resolve the conflict is to let the lower priority transaction execute at a higher priority until it has released the needed resources. In the simplest form, the lower priority transaction inherits the priority of the higher priority transaction. This method is called Priority Inheritance, as proposed in [SRL90]. This can lead to a priority inheritance chain where several low priority transactions must gain a higher priority and finish execution before the high priority transaction gets the requested resources. In Priority Ceiling, also proposed in [SRL90], the high priority transaction is never blocked for more than a single lower priority transaction. This is achieved by keeping a priority ceiling to all locked resources and keeping an ordered list of the prioritized transactions. A transaction may start a new critical section only if its priority is higher than the ceiling priorities of all the resources in the section. The algorithm needs fixed priorities to the transactions. A similar method called Dynamic Priority Ceiling is proposed in [SRSC91], except that it accepts priorities that are relative to transaction deadlines while Priority Ceiling needs fixed resources.

Two-phase locking (2PL) [EGLT76] and its variants are the most common pessimistic concurrency control method in current database management systems. They perform well in traditional databases without transaction deadlines. Concurrency control in 2PL and its variants is based on locks. The locks are used in two phases: expanding phase and shrinking phase. In the expanding phase a transaction asks locks to the resources it wants to access. In the shrinking phase the transaction releases the accessed locks and locked resources. The transaction is allowed to ask for more locks as long as it has not released any of the previous ones.

Optimistic methods are suitable for real-time databases. It is often easier to let a transaction execute up to the commit point and then do a conflict check than to have full bookkeeping of hard locks in pessimistic methods. Moreover, neither deadlocks nor priority inversion happen in optimistic methods. Optimistic methods are not used in traditional databases since every restart of a transaction is expensive. The situation is different for real-time databases where the value of a transaction is nullified or lowered after the deadline. Hence, if a low-priority transaction is aborted due to a conflict, it may not need a restart at all if it has already missed its deadline or has no way to reach it. Naturally the number of aborted transactions should be minimized since resources are still wasted. Indeed, if deadlines allow, it may sometimes be reasonable to let a low priority transaction commit and abort higher priority transactions. Usually at most two transactions are conflicting with the committing transaction. The committing transaction is known to commit while none of the conflicting transactions are guaranteed to meet deadlines and commit.

The original optimistic methods are all used in real-time databases, but new methods have also been designed. Perhaps the best method is called Dynamic adjustment of serialization order, or OCC-TI [LS93]. It is based on the idea that transactions can be ordered dynamically at the validation phase to minimize conflicts and hence restarts. A revised version of the method that further minimizes the number of restarts has been presented by Lindström [Lin00].

A very interesting family of concurrency control algorithms, called Speculative Concurrency Control (SCC) algorithms, is the latest major addition to the concurrency control field [BB95, BB96, BN96]. The SCC algorithms define a set of alternate transaction schedulings when a conflict candidate is detected. These *shadow* transactions execute speculatively on behalf of a given uncommitted transaction to protect against blockages and restarts The shadow transactions are adopted only if one or more of the suspected conflicts materialize. In the generic Speculative Concurrency Control algorithm with k Shadows (SCC-kS), up to k shadows are generated after conflict candidate detection.

The SCC-kS algorithm shows a very interesting approach to concurrency control. It uses spare resources in a DBMS to create the shadow transactions. The authors have also proposed a RT-DBMS variation with deferred commit called SCC-DC that considers transaction deadlines and criticality [BB96].

The question of what concurrency control methods best suit a database management system at different workloads and environments has been a major theme in performance studies and concurrency control research, both in traditional and real-time databases. In conventional database systems pessimistic algorithms that detect conflicts before data item access and resolve them by blocking transactions give better performance than optimistic algorithms, especially when physical resources are limited. If resource utilization is low enough so that even a large amount of wasted resources can be tolerated, and there are a large number of transactions available to execute, then a restart-oriented optimistic method is a better choice [BHG87]. In addition, the delayed conflict resolution of the optimistic approach helps in making better decisions in conflict resolution, since more information about conflicting transactions is available at the later stage [LS94]. On the other hand, the immediate conflict resolution policy of pessimistic methods may lead to useless blocking and restarts in real-time databases due to its lack of information on conflicting transactions at the time when a conflict is detected [HCL92]. Also the SCC approach is interesting when resources are sufficient.

2.3   Distributed databases

The history of distributed databases goes back to times when both disks and memory were expensive. Since those times the prices of memory and disks have decreased enough that the advantages of distribution compared to a single system are not significant. Lately, however, new distributed applications find a distributed database architecture useful. Such applications include multimedia, World Wide Web, and telecommunications.

The authors in [CP84] give a definition for a distributed database:

> A distributed database is a collection of data which are distributed over different
> computers of a computer network. Each site of the network has autonomous pro-
> cessing capability and can perform local applications. Each site also participates
> in the execution of at least one global application, which requires accessing data at
> several sites using a communication subsystem.

Data in a distributed database is divided into fragments. Each fragment is a set of data items
that are managed by a computer, or rather a local database management system in the computer.
From now on we call the pair computer and its database management system *a node*. A node
in a distributed database may include several fragments, and a fragment may belong to several
nodes. Usually it is preferred that a node holds exactly one fragment of data, and a fragment
belongs to exactly one node. It simplifies distributed database management and cooperation
between database nodes.

Since the definition of a distributed database includes a set of local applications, or databases,
it is natural to compare distributed and traditional non-distributed databases. It turns out that
the distributed database is very close to a traditional database. This implies that traditional
database results can be used or modified to suit distribution. The areas to consider are the
ACID-properties, data independence, data redundancy, physical implementation, recovery, and
concurrency control.

- *ACID-properties.* The ACID-properties, as mentioned earlier, are atomicity, consistency,
  isolation, and durability. We will cover them in turn here.

  - *Atomicity.* Both traditional and distributed transactions are atomic. That is, the
    writes that a transaction executes are visible to other transactions only after the
    transaction commits. However, the methods to achieve atomicity differ in imple-
    mentations. In a distributed environment communication between nodes must also
    be taken into account.

  - *Consistency.* Data in both a traditional database and in a distributed database must
    be consistent. The consistency to committed transactions is important since data can

be internally inconsistent as long as it is visible only to an uncommitted transaction. The same is true in a distributed database. The database must remain consistent also in case of communication failures between nodes.

– *Isolation.* An uncommitted transaction does not affect calculations and results of other transactions. In a distributed database this includes both local-global and global-global transaction isolation. The former occurs when a local and a distributed transaction try to access the same resources. The latter occurs when two distributed transactions compete about the same resources.

– *Durability.* Once a transaction commits its changes to the data become permanent in the database. This holds true both in a traditional and in a distributed database. If the changes affect a replicated data item, all replicants should hold the same value after the commit, or alternatively all replicants with old values should be made invalid.

• *Data independence.* Data independence states that the stored data in a database is independent of the actual implementation of the data structures. This is true both for traditional and distributed databases. However, distributed databases have a new aspect: distribution transparency. It states that to a certain level data is independent of its location. At the highest level a transaction does not see data distribution at all. At the lowest level each application can see data distribution and access data in a node independently from other nodes and other applications.

• *Data redundancy.* In a traditional database, redundant data is considered a drawback to the system performance. Due to this, various approaches, such as normal forms in relational databases, have been designed to reduce data redundancy. In a distributed database, data can sometimes be redundant. The level of redundancy depends on the chosen design approach. Usually a single node holds one data item only once. A multiversioning database is an exception to this rule. It may hold several versions of the same data item. Multiversioning is used in traditional databases as well. On the other hand, some data may have copies on different nodes. This is called data replication. Usually replication is invisible to transactions. It is used for minimizing distributed read transactions. The drawback is the extra complexity of update transactions and fault-tolerance.

- *Fault tolerance and recovery.* In a traditional database, fault-tolerance and recovery includes policies for aborted transactions, and for physical and software system failures. When data is committed, it is available to other transactions. In a fault-tolerant environment, the recovery procedure is invisible to the users. The database is available even during a system failure. In a distributed environment, also site and network failures must be taken into account. This affects situations when nodes must co-operate and exchange messages. It is not possible to create a protocol that can handle all recovery situations in a distributed database [BHG87].

- *Concurrency control.* Concurrency control allows transactions to execute in parallel and still see the database as a single entity that is only for their use. This is achieved via a concurrency control policy. The chosen policy must guarantee that transaction execution maintains data consistency. In a distributed database, two levels of concurrency control must be taken into account. First, the local concurrency control policy in a node is similar to the policy in a traditional database. Second, a global concurrency control policy is needed to guarantee serializability between distributed transactions. A local concurrency control policy is not sufficient for this.

The concurrency control policy in a distributed database may be divided into two policies: local policy and global policy. The local policy affects all transactions that are active in a node. The global policy affects only transactions that are active in several nodes.

When a transaction exists only in a single node, it may still affect the global concurrency control. For instance, let us have transactions T_1, T_2, and T_3 in the distributed database. The transactions T_1 and T_3 are active in nodes 1 and 2, while transaction T_2 is active only in node 1. Moreover let us have the following histories:

$$H_1 = r_1[x]w_2[x]r_2[y]w_3[y]c_1c_2c_3,$$

and

$$H_2 = r_3[z]w_1[z]c_3c_1.$$

If we take only global transactions T_1 and T_3, we do not have a cycle in the global serialization graph $SG(H_1 \cup H_2)$. However, when T_2 is included, we have a cycle $T_1 \rightarrow T_2 \rightarrow T_3 \rightarrow T_1$ which needs to be detected.

A local concurrency control policy is similar to the local policy in non-distributed databases. Each transaction is controlled by the local concurrency controller. In principle, such transactions could have a different concurrency control policy on different nodes.

A global concurrency control policy needs to take care of conflicts that affect several nodes. In order to achieve this, three alternate policies are possible.

1. Local concurrency controllers co-operate to achieve global concurrency control. This is seldom done since depending on the chosen global concurrency control policy this can generate several messaging phases between local nodes. The overhead of such a distributed concurrency control policy is too high for most cases.

2. The distributed database has a global concurrency control coordinator. The coordinator is responsible for the global concurrency control policy. The coordinator takes care of the global serialization by keeping track of global transaction conflicts. The advantage of this approach is that local concurrency controllers need not be aware of the global concurrency control policy. The global coordinator can manage all conflicts alone as long as it informs the local nodes of what to do in conflicts. The drawback of the approach is that the global coordinator can easily become a bottleneck of the system. Nevertheless this approach is better than the democratic approach where local concurrency controllers co-operate.

3. All transactions are considered local, and global concurrency control is achieved by controlling transactions in nodes. This is a restricting approach that works only with few concurrency control mechanisms.

When a new transaction request arrives to a transaction manager, it creates a new local transaction for it. Alternatively, when it is known which nodes will participate in this transaction, the transaction may immediately be divided into a set of subtransactions, each of which executes in one node. The transaction in the arriving node then becomes the leader of this transaction.

A local transaction may be promoted to a distributed transaction when it wants to access a data item that is not in its node. In such a case the transaction manager sends a subtransaction request to the node where the requested data item resides. Due to this the nodes must have knowledge of data in other nodes. Usually this is implemented with directories. The leader transaction gathers collected information from the subtransactions. When all necessary information is collected, the leader is ready to commit.

Once the leader is ready to commit, it co-operates with the subtransactions to find out if the transaction should commit or abort. There are several algorithms to implement cooperation, but the basic idea in all of them is that the commit should be atomic. Either the leader and subtransactions are all committed, or they are all aborted. This is a approach that never violates consistency.

In any atomic commit protocol, each process may cast one of two votes for global commit: Yes or No, and may reach one of two decisions: commit or abort. Furthermore, in order to guarantee atomicity, the atomic commit protocol must allow all participants to reach decisions that guarantee the following rules [BHG87]:

1. All participants that reach a decision reach the same one.

2. A participant cannot reverse its decision once it has reached one.

3. The commit decision can only be reached if all participants voted Yes.

4. If there are no failures and all processes voted Yes, then the decision will be to commit.

5. Consider any execution containing only failures that the algorithm is designed to tolerate. At any point in this execution, if all existing failures are repaired and no new failures occur for sufficiently long, then all processes will eventually reach a decision.

Various protocols have been introduced to preserve atomicity. The best known of these is called the two-phase commit (2PC) protocol [Gra78]. Assuming no failures, it goes as follows:

1. The coordinator sends a VOTE-REQ (vote request) message to all subtransactions.

2. When a subtransaction receives a VOTE-REQ, it responds by sending to the coordinator a message containing that participant's vote: Yes or No. If the subtransaction votes No, it decides to abort and stops.

3. The coordinator collects the vote messages from all participants. If all of them were Yes and the coordinator's vote is also Yes, the coordinator decides to commit and sends COMMIT messages to all participants. Otherwise, the coordinator decides to abort and sends ABORT messages to all participants that voted Yes. In either case, the coordinator then stops.

4. Each subtransaction that voted Yes waits for a COMMIT or ABORT message from the coordinator. When it receives the message, it decides accordingly and stops.

The basic 2PC satisfies the listed rules for atomic commit protocols as long as failures do not occur. However, when for some reason the communication between subtransactions and the coordinator is distracted, it is possible that a subtransaction is uncertain of the correct behavior. In such a case the subtransaction is in a blocked state. It neither can abort or commit since it does not know what the other subtransactions and the coordinator have decided. Naturally the blocked subtransaction may wait as long as the failure is corrected. This satisfies requirement 5 earlier. Unfortunately the decision may be blocked for an indefinitely long time.

Since the uncertainty of the correct decision can cause serious problems, it would be desirable to have an atomic commit protocol where participants do not have uncertainty times. Unfortunately such a protocol would require that a participant would cast a vote and learn the votes of all other participants all at once. In general, this is not possible [BHG87].

The basic 2PC may be too strict in a real-time environment. Time may be wasted in the commit procedure when ready subtransactions must block their updated data items when they are waiting for the global commit result from the leader. Due to this, a protocol called The Real-time Implicit Yes-Vote protocol was proposed 1996 by Al-Houmaily and Chrysanthis [AHC96]. The protocol is based on the assumptions that 1) each node employs a strict 2PL protocol for concurrency control that takes into consideration the priority of transactions, and 2) a specific distributed logging protocol called physical page-level write-ahead logging. With these assumptions the protocol uses commit processing that is overlapped with the execution of the transactions' operations [AHC96]. This algorithm is better suited for real-time environments than a regular two-phase commit since it takes the priorities into account and allows transactions to execute in parallel with the commit processing.

An optimistic 2PC-based commit protocol PROMPT (Permits Reading Of Modified Prepared-data for Timeliness) is based on the assumption that a distributed transaction will not be aborted at commit time. The committing transaction can lend data to other transactions so that it does not block them. In the algorithm, two situations may arise depending on the finishing times of the committing transactions [GHRS96, HRG00]:

- *Lender Finishes First.* In this case the lending transaction receives its global decision before the borrowing transaction. If the global decision is to commit, both transactions are allowed to commit. If the decision is to abort, both transactions are aborted. The lender is naturally aborted because of the abort decision. The borrower is aborted because it has read inconsistent data.

- *Borrower Finishes First.* In this case the borrower has reached its committing stage before the lender. The borrower is now made to wait and not allowed to send a YES vote in response to the coordinator's PREPARE message. The borrower has to wait until such time as the lender receives its global decision or its own deadline expires, whichever comes earlier. In the former case, if the lender commits, the borrower is allowed to respond to the coordinator's message. In the latter case, the borrower is aborted since it has read inconsistent data.

In summary, the protocol allows transactions to read uncommitted data in the optimistic belief that the lower priority transaction that is holding data will eventually commit. Usually it is not recommended that a transaction reads uncommitted data because this can lead to the problem of cascading aborts [BHG87]. However, the problem does not occur in Optimistic Commit Protocol because the abort chain includes only the lender and the borrower. Other transactions are not aborted if the lender aborts because the borrower is not in the prepared state and hence it does not cause any further aborts [GHRS96].

An interesting class of atomic commit protocols called one-phase commit protocols have got more attention especially in distributed real-time database research [AGP98], [HR00, SC90]. There is no explicit voting phase to decide on the outcome (commit or abort) of the transaction in these protocols since subtransactions enter the prepared state at the time of sending the work completion message itself. Thus, the commit processing and data processing activities are overlapped since not all subtransactions enter the prepare state at the same time. This in turn causes longer blocking times since the prepared subtransactions cannot be aborted before final consensus is reached. Due to this one-phase commit protocols are best suited for distributed transactions with small subtransactions.

2.4   Intelligent Networks

Intelligent Networks (IN) are the first step towards an open telecommunications service environment. The basic idea of IN is to move service logic from switches to special IN function logics called Service control functions (SCF). With SCFs it is possible to let telephone switching software be relatively static when new service logic programs are used in SCFs to implement new services. Hence IN architecture can be used for special service creation, testing, maintenance, and execution.

Currently IN is used in most telecommunications environments. It is under standardization both in Europe and in the U.S. The standardization committees, ITU-T (International Telecommunication Union - Telecommunication standardization sector), ETSI (European Telecommunications Standards Institute), and ANSI AIN committee (American National Standards Institute Advanced Intelligent Network committee) are all defining the IN concept. ETSI and ITU-T define practically equal IN concepts. ANSI AIN is somewhat different although there is substantial co-operation with the European standardization committees. We follow the ITU-T/ETSI recommendations for the Intelligent Networks.

The basic standard that defines the framework of other IN standards in ITU-T is the Q.120x series. It defines a framework for ITU-T IN standards with possible revisions. The standards themselves are structured to Capability sets. The first Capability set CS-1 was introduced in the

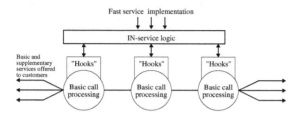

Figure 2.1: Q.1201/IN service processing model.

Q.121x series. The future Capability sets will be introduced in the Q.12nx series, where n is the capability set number. This Section is based on IN CS-1 in the Q.121x series. The Capability sets are backward compatible to the previous ones so that the implementation of services can be progressed through a sequence of phases [GRKK93].

The CS-1 capabilities support IN services that apply to only one party in a call. The CS-1 capability set is independent of both the service and topology levels to any other call parties [ITU93b]. In other words, the services are only dependent on a single call party and the same service may be used again on a different environment with possibly different parameters. This definition restricts the CS-1 service definitions, but in turn it gives a good basis for the future IN architectures.

The IN environment is based on hooks that recognize IN-based services. When a service request arrives, a trigger in a basic call processing node activates the actual IN service logic (Figure 2.1). Once the IN service is finished, the basic call processing gains control again. The triggered IN service program consists of a set of Service independent building blocks (SIBs).

New IN services may be implemented from IN components with hooks and SIBs. The switch software uses hooks to call IN service logic when a special service occurs and the switch itself remains stable in time. The actual service implementation is left to the network main-tainer. The IN conceptual model defines the different IN components that are used for service creation, use, and maintenance. The conceptual model also defines the CS-1 standardized ser-vices, service features and service-independent building blocks.

2.4.1   IN Conceptual Model

The IN conceptual model is divided into four planes that define different views to the IN model. The planes, as can be seen in Figure 2.2, are the Service plane that defines IN services and service features [CCI92b], the Global functional plane that defines how service features are implemented by SIBs [CCI92a], the Distributed functional plane that defines how operations that SIBs need are mapped to functional entities [IT93a], and the Physical plane that defines network entities that encapsulate functional entities [IT93b].

Service plane

The visible items in the Service plane are the IN services. A service consists of a set of service features that build the service. The service plane represents an exclusively service-oriented view that does not contain any information of service implementation aspects. The architecture that is used for the implementation is invisible to the Service plane.

The IN CS-1 defines a set of services and service features that are visible to the Service plane. The services are defined in the CS-1 to describe the general IN capabilities. The actual implemented set of services may be smaller than what is listed in CS-1. The services are structured from service features. A service feature is a specific aspect of a service that can be used with other service features to implement new services.

Global functional plane

The Global functional plane defines service feature functionality which is achieved by Service-independent building blocks (SIBs). Services that are identified in the service plane are decomposed into their service features, and then mapped into one or more SIBs in the Global functional plane. The Basic call process (BCP) functional entity is the only entity that is visible to the service. The Point of initiation (POI) initiates IN service logic and the Point of return (POR) returns control back to the basic call process.

The main SIB for database services is the Service data management SIB. It enables user-specific data to be replaced, retrieved, incremented, or decremented. It is the main persistent

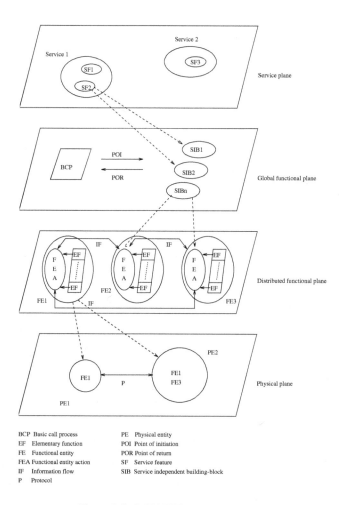

BCP Basic call process
EF   Elementary function
FE   Functional entity
FEA Functional entity action
IF   Information flow
P   Protocol

PE   Physical entity
POI Point of initiation
POR Point of return
SF   Service feature
SIB Service independent building-block

Figure 2.2: Q.1201/IN conceptual model.

data management block [ITU93a]. The other database-related SIB is the Translate SIB. It translates input information and provides output information. It can be used for number translations
that may depend on other parameters such as time of a day [ITU93a].

Distributed functional plane

The Distributed functional plane defines functional entities that SIBs use for actual service
implementation. A functional entity is a group of functions in a single location and a subset
of the total set of functions that is required to provide a service. Every functional entity is
responsible for a certain action. Together they form the IN service logic. The entities are visible
only on the Distributed functional plane which defines the logical functional model of the IN
network. The physical network entities consist of functional entities. They are defined in the
Physical plane.

The functional entities and their relationships that CS-1 defines are in Figure 2.3. The set of
relationships in the IN distributed functional plane model also defines the regular IN call profile
by showing how information flows between the functional entities. The first entity to reach is
the Call control agent function (CCAF). It is an interface between users and network call control
functions. It is first accessed when a user initiates a connection and gets a dial tone. After that
the call control continues to the Call control function (CCF). It performs regular switching by
connecting and serving the call. In regular calls the only functional entities that are accessed
are CCAF and CCF.

The Service switching function (SSF) recognizes service control triggering conditions when
a special IN service is needed. It controls the IN service by co-operating with the Service control
function (SCF). SCF is the main functional entity of the IN distributed functional plane model.
It executes service logic programs that consist of SIBs. When the service is finished the SCF
returns control to the SSF. It may also communicate with several SSFs depending on the nature
of the service. For instance, SCF may consult both the caller SSF and the called SSF.

In service logic program execution, the SCF may also interact with other functional entities
in order to access additional logic or to obtain information that is required to process a service
request. The main entities that the SCF communicates with are the Special resource function

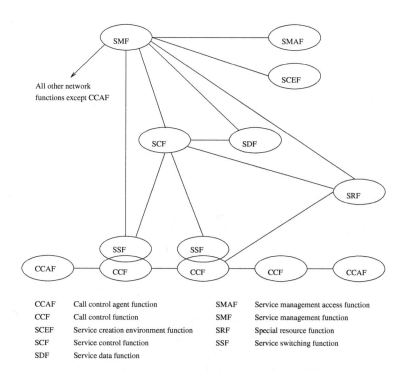

CCAF	Call control agent function	SMAF	Service management access function
CCF	Call control function	SMF	Service management function
SCEF	Service creation environment function	SRF	Special resource function
SCF	Service control function	SSF	Service switching function
SDF	Service data function		

Figure 2.3: Q.1204/IN distributed functional plane model.

(SRF) and the Service data function (SDF).

The SRF provides specialized resources that are required for service execution, such as voice recognition, digit receivers, announcements, and conference bridges. The exact resources that the SRF offers are not specified in the CS-1. It is left open for any services that need specialized resources. The SCF gives control to the SRF during the special resource execution. The SRF receives results from the SCF when the execution is finished.

The service, customer, and network related data is in a set of SDFs. The SCF communicates with one or more SDFs to gain customer and network data that is needed for the executing service. The SDFs offer most of the database services for the IN functional architecture. Again, the implementation of a SDF architecture is not specified in CS-1.

Finally, when the IN service is executed, the SCF returns control to the SSF which returns it to the CCF. Once control is returned to the CCF, the CCF continues normal call control execution, if necessary. Often the IN-triggering call is terminated when the SCF returns control to the SSF since most IN-specific services need access to SCF and IN-specific data during the complete call.

The rest of the functional entities, Service management function (SMF), Service management access function (SMAF), and Service creation environment function (SCEF), form a service maintenance environment where the service maintainers can design, implement, and maintain IN services and service features. The Service management access function is the entry function to reach the management services. It provides an interface between service managers and the service managing environment. The SCEF provides an environment where new services can be defined, developed, tested, and input to the management environment.

The main component of the management is the Service management function (SMF). It is used for all IN functional entity and service maintenance. It can also be used for gathering statistic information or for modifications of service data. The management environment is undefined in CS-1, but it will probably be closely related to the Telecommunications Management Network architecture (TMN).

Physical plane

The lowest plane in the IN conceptual model is the Physical plane. It describes how the functional entities are mapped into physical entities, and how the network itself is implemented. The
physical architecture that is recommended in Q.1215 is in Figure 2.4, but the physical plane architecture is not strictly defined. It is possible to have various different mappings from logical
entities to physical entities.

The main physical entry point to the network is the Network access point (NAP). It consists
of a CCF and possibly a CCAF that allow the connection. The connection is then transported to
the Service switching point (SSP), which is the main entity of the IN physical architecture. The
SSP implements switching functions and IN service triggering. It consists of functional entities
CCF and SSF. It may also consist of a SCF, a SRF, a CCAF, and a SDF. The definition allows a
physical architecture that consists of a SSP and the management entities alone.

Other versions of the SSP are the Service switching and control point (SSCP) and the Service node (SN). A SSCP may have the same functional entities as the SSP but it has a SCF
and a SDF as core entities. A SN has a CCF, a SSF, a SCF, a SDF, and a SRF as core entities.
In practice it is possible to add most functional entities to a single physical entity, and call it
accordingly. The definition allows different distribution levels to the architecture. There are test
architectures from total centralization to total distribution.

The presented architecture is an example of how the physical plane can be implemented.
The recommendations give exact specifications of mappings from functional entities to physical entities. The functional entities that implement IN logic are more important than their
distribution to physical entities.

The actual IN service logic processing occurs in a Service control point (SCP). Functionally
it consists of a SCF and optionally a SDF. The SCP contains service logic programs and data that
are used to provide IN services. Multiple SCPs may contain the same service logic programs
and data to improve service reliability and load sharing. The SCP may access data directly
from a SDP or through the signaling network. The SCP may also access data from a SDP in an
external network. The SDP is responsible for giving a reply which either answers the request

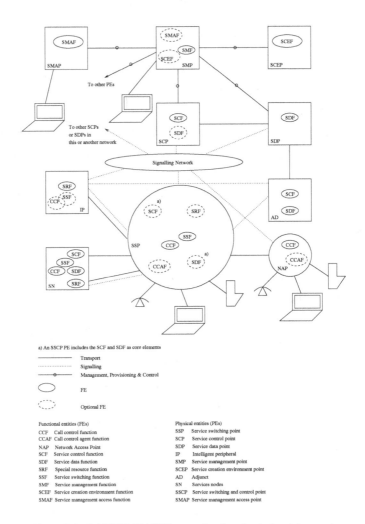

a) An SSCP PE includes the SCF and SDF as core elements

⎯⎯⎯⎯⎯⎯   Transport
....   Signalling
⎯⎯o⎯⎯   Management, Provisioning & Control

⬭   FE

⬭   Optional FE

Functional entities (FEs)		Physical entities (PEs)	
CCF	Call control function	SSP	Service switching point
CCAF	Call control agent function	SCP	Service control point
NAP	Network Access Point	SDP	Service data point
SCF	Service control function	IP	Intelligent peripheral
SDF	Service data function	SMP	Service management point
SRF	Special resource function	SCEP	Service creation environment point
SSF	Service switching function	AD	Adjunct
SMF	Service management function	SN	Services nodes
SCEF	Service creation environment function	SSCP	Service switching and control point
SMAF	Service management access function	SMAP	Service management access point

Figure 2.4: Q.1205/Q.1215/IN scenarios for physical architectures.

or gives an address of another SDP that can answer it. An Adjunct (AD) is a special SCP that has both a SCF and a SDF as core entities. As in switching points, various functional entity distributions allow different distribution levels in the service control and data points.

The main physical database entity is the Service data point (SDP). It contains data that the service logic programs need for individualized services. It can be accessed directly from a SCP or a SMP, or through the signaling network. It may also request services from other SDPs in its own or other networks. Functionally it contains a SDF which offers database functionality and real-time access to data.

The Intelligent peripheral (IP) is responsible for processing special IN requests such as speech recognition or call conference bridges. Functionally it consists of a SRF, and possibly a SSF and a CCF. Usually it does not contain any entities other than the SRF.

Finally, the IN management is left to the Service management point (SMP) together with the Service creation environment point (SCEP) and Service management access point (SMAP). A SMAP is used for maintenance access to the management points. Functionally it consists of a SMAF. The SCEP implements a service creation and testing platform. Functionally it consist of a SCEF. The SMP handles the actual service management. It performs service management control, service provision control, and service deployment control. It manages the other IN physical entities, also the SDPs. It consists of at least a SMF, but optionally it may also consist of a SMAF. Again, this allows different functional distribution levels to the implementation.

2.4.2   The Future of Intelligent Networks

The IN architecture, as described in CS-1, is already several years old. Naturally IN standardization has evolved a lot since the times of CS-1. A number of groups are currently developing technologies aimed at evolving and enhancing the capabilities of Intelligent Networks. These initiatives include project PINT on how Internet applications can request and enrich telecommunications services, The Parlay consortium that specifies an object-oriented service control application interface for IN services, and the IN/CORBA interworking specification that enables CORBA-based systems to interwork with an existing IN infrastructure [BJMC00]. Similarly Telecommunications Information Networking Architecture (TINA) has been under heavy development and research since 1993. TINA concepts are easily adaptable to the evolution of IN [MC00].

The overall trend in IN seems to go towards IP-based protocols and easy integration to Internet services. For instance, Finkelstein et al. examine how IN can interwork with the Internet and packet-based networks to produce new hybrid services [FGSW00]. However, the underlying IN architecture is not changing in these scenarios. Hence the analysis and descriptions we have given here are relevant regardless of future trends. The interworking services are at lower abstraction levels, and need new gateways and interfaces to connect IN and IP-based services. This in turn affects the interfaces to globally accessible IN elements such as SSPs and perhaps SCFs. Even SDFs may need to have an interface for CORBA or TINA based requests.

2.5   Current and future mobile networks

Mobile networks use radio techniques to let portable communication devices (usually telephones) connect to telecommunications networks. The number of mobile connections is growing rapidly while the number of connections in fixed networks has remained relatively stable. If the trend continues, mobile networks will be a leading factor in the competition of teleoperators.

Currently first (1G) and second (2G) generation mobile systems are used in telecommunications. The 1G mobile systems are based on analog techniques. Such networks are Nordic Mobile Telephone (NMT) in the Nordic countries and Advanced Mobile Telephone System (AMPS) in the U.S. The 1G systems no longer have a strong impact on mobile network architectures due to their analog nature, voice-alone service, and closed architectures.

The 2G mobile networks, such as Global System for Mobile communications (GSM), use digital techniques for mobility implementation. Their functional architectures are well designed and applicable for more than pure mobile connection services. The digital architectures allow separation of services and connections.

The 2G architectures have a completely different approach from analog and closed 1G architectures. In 2G design defining open interfaces, and digital data and speech encoding are the most important aspects. The data rate of 9.6kb/s was reasonable at the time of 2G definition. It was not a surprise that 2G soon became a major success, and especially GSM spread almost everywhere in the world.

2.5.1 GSM architecture

The GSM architecture is built on a Public land mobile network (PLMN). It relies on fixed networks in routing, except between a Mobile station (MS) and the GSM network. The network has three major subsystems: the Base station subsystem (BSS), the Network and switching subsystem (NSS), and the Operation subsystem (OSS) [MP92].

Next to the subsystems BSS, NSS, and OSS, the GSM network consists of a set of MSs that usually are the only elements visible to the end users. The service area is divided into a number of cells that a BSS serves. A MS has radio communications capability to the base station of that cell. The Mobile service switching center (MSC) provides switching functions and interconnects the MS with other MSs through the BSS or co-operates with public fixed networks [Jab92].

The BSS includes the machines in charge of transmission and reception to the MSs. It controls communication between MSs and the NSS which is responsible for routing calls from and to MSs. The BSS consist of a set of Base transceiver stations (BTSs) that are in contact with MSs, and a set of Base station controllers (BSCs) that are in contact with BTSs and the NSS. The BTS manages GSM cells that define coverage for MSs. The BSC manages the radio resources for one or more BTSs.

The NSS includes the main switching functions of GSM and the necessary databases that are needed for subscriber data and mobility management. The name Network and switching subsystem comes from Mouly and Pautet [MP92]. The subsystem is also called the Switching subsystem since a GSM network includes both a BSS and a NSS [MP92]. The main role of the NSS is to manage communications between GSM users and other telecommunications network users.

The elements of the NSS can be seen in Figure 2.5. Next to the elements, the figure also

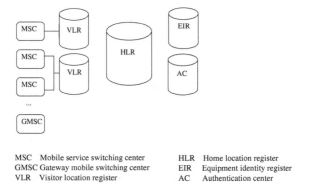

MSC   Mobile service switching center   HLR   Home location register
GMSC Gateway mobile switching center   EIR   Equipment identity register
VLR   Visitor location register   AC   Authentication center

Figure 2.5: GSM Network and Switching Subsystem elements.

shows a typical configuration of a NSS in a single teleoperator. The NSS includes a set of MSCs
that route MS calls, a set of Visitor location registers (VLRs) related to one or more MSCs that
hold location information of mobile stations within its region, one or more Gateway mobile
service switching centers (GMSC) that are responsible for routing calls to external networks, an
Equipment identity register (EIR) that maintains information for identifying Mobile equipment,
and an Authentication center (AC) that is responsible for management of security data for the
authentication of subscribers.

Information on the location of the MS is updated in the reference database HLR. It contains
relevant permanent and temporary data about each subscriber. Additionally, information is
maintained in a local VLR. The VLR, corresponding to one or several MSCs, contains detailed
data of MS locations and subscribed GSM services. The MS-specific data is used for routing
incoming and outgoing calls. VLR functions cover location updates, temporary mobile station
identity allocation for location confidentiality, and storage of a MS roaming number (MSRN)
[Jab92]. The HLR is relatively static, while data in the VLR changes when the subscriber
moves. The Signaling system 7 (SS7) connects MSCs, VLRs, and HLRs [PMHG95].

The OSS has three main tasks: subscriber management, system maintenance, and system
engineering and operation [MP92]. The OSS architecture is designed so that it is long-term
compatible with Telecommunications management network (TMN) [Pon93]. The TMN itself

is a complex and well defined architecture for telecommunications management. We do not cover TMN architecture in this thesis.

Subscriber management in the OSS has two main tasks: subscription and billing. The subscription task occurs when a new subscriber joins to the network. The information of the subscriber and her mobile equipments must be registered. This involves the HLR that holds the subscriber information, and possible other service provider related databases that are not listed in the GSM standard. The standard does not restrict the use of such databases [MP92]. The billing task occurs when a subscriber is billed from a service use. Usually the triggering client of the service is a call, but it can also be a supplementary service or some external teleoperator of an incoming call in case the subscriber is located in an external PLMN.

System maintenance includes the failure detection and repair of all GSM components. This is often accomplished by monitor computers and alarms.

The System Engineering and Operation is closely related to System Maintenance. However, in this category the management staff is more concerned of what kind of network architecture to create and how to run it efficiently.

2.5.2   Third generation mobile networks and beyond

Currently the third generation mobile networks (3G) are under development. They are based on second generation architectures but they also rely on IN services for multimedia architecture. Such systems are the Future Public Land Mobile Telecommunication System (FPLMTS) that is under standardization in ITU Radiocommunications sector (ITU-R) and in ITU-T, and Universal Mobile Telecommunication System (UMTS) that is under standardization in ETSI. The UMTS is closely aligned to the FPLMTS. It is expected that the UMTS and the FPLMTS will be compatible in the long run. Since the definition of the 3G wireless networks, network system architectures have gone through evolutionary phases and now the first 3G wireless network operators are ready to offer services in America, Europe, and Asia.

Some of the system characteristics for the FPLMTS were defined in the mid-90s [Pan95]: a digital system using 1.8 – 2.2 GHz band, multiple radio environments (cellular, cordless, satellites, and fixed wireless), multimode terminals to provide roaming capability, a wide range of telecommunication services, high quality and integrity – comparable to fixed networks, international roaming and inter system handover capability, use of IN capabilities for mobility management and service control, high levels of security and privacy, and flexible and open network architectures.

The UMTS provides wireless communication services to users on the move. The services will include high speed data services and video services like mobile videophone [Mit95]. The users will be provided with data rates of up to 144 kbit/s in macrocellular environments, up to 384 kbit/s in microcellular environments, and up to 2 Mbit/s in indoor or picocellular environments. The increase of data rate is the main benefit of UMTS systems. With higher data rates new multimedia services may be implemented.

The main functions of an UMTS database service are as follows [Mit95]:

- It offers basic database management system functionality of providing read and write access to data,

- it locates requested data within the database,

- it moves data to locations close to the user when the user has roamed to provide fast access to data where it is frequently being used,

- it provides mechanisms that ensure the integrity and reliability of data even when it is copied between locations and possibly duplicated, and

- it ensures that data is only accessed by entities that have been authorized to do so.

An interesting fact about database services in the UMTS is that their implementation is left undefined in UMTS standardization. The radio interface independent functions, essentially call control and mobility management, are outside the scope of the specifications and are handled by the core network [Ric00]. It is important to draw a distinction between 3G networks and 3G services.

Open interfaces are defined that allow teleoperators to use existing network architectures as much as possible. In practice this implies that UMTS mobility and call control support at architecture level will be similar to 2G mobility and call control support. New special teleoperator-specific services will be implemented on top of the network and thus left undefined in the standardization.

According to Pandya [Pan95] and Mitts [Mit95] Intelligent Networks and SS7 are seen as the basis for UMTS service implementation. Since the mid-90s when those articles were written, UMTS has evolved towards IP-based implementations. Yet IN still has a lot to offer to 3G networks. In principle, all 3G database services could be left for SDFs and SMFs. This includes HLR, VLR, and EIR databases. Even when 3G architectures are fully IP-based, the IN architecture may be used on top of the IP-network.

Special services, management, and mobility need very different types of database services. Special services need fast read access to data, management needs reliable distributed updates, and mobility needs small and fast updates with the cost of reliability. It is imminent that not all three can be fully satisfied. Compromises are needed if a single database management system is intended to support all the categories.

The 3G wireless networks are not the final solution to wireless networks. First, they are based on old ATM techniques. Second, their throughput is limited to 2Mb/s which is not enough for modern multimedia services [DF01]. Due to the limitations in 3G networks, the next evolution to 4G wireless networks is under research. According to Becchetti et al. the 4G networks will not be based on a standard interface in the same way as the 3G networks [BPI$^+$01]. While it is too early to state anything specific, it is clear that 4G wireless networks will be specialized on various application domains, standard interfaces between them, transparency, and quality of service. The architectures are IP-based rather than ATM-based [RKV$^+$01, TNA$^+$01].

The benefits of a fully IP-based architecture are clear. The IP approach allows a wide range of new services and shortens the service creation cycle. For example, voice/multimedia calls can be integrated with other services [YK00]. The cost on service creation, provisioning, and maintenance can be reduced due to a common platform. Also cost savings on ownership, management, transport, and interface implementation are significant.

It should be noted that while 3G is starting and 4G is under research, the limits of 2G techniques are not yet met. The Wireless Application Protocol (WAP), General Packet Radio

Service (GPRS), and enhanced data rates for GSM evolution (EDGE) have evolved GSM to new usage levels [Pra99]. It is expected that GSM will dominate the mobile communication market long after the first 3G network implementations are commercially available [Kon00]. Yet eventually 2G wireless networks will evolve to 3G wireless networks, and finally to full IP-based 4G wireless networks. We expect 2G wireless networks to be obsolete in 10 years, and 3G wireless networks in 15-20 years.

Chapter 3

IN/GSM data analysis

As we have already seen in the previous section, in the future it is most probable that IN will support mobility, and mobile networks will use IN services as a basis for special services. Due to this it is natural to consider a common database management system architecture for IN and GSM. We call this an IN/GSM database architecture.

In order to define the IN/GSM database management system architecture we must first understand the special nature of IN and GSM data and data stores. Both platforms offer a rich set of details that can be analyzed and thus maintained in the IN/GSM database architecture.

The most important aspect in an IN/GSM database architecture is to support real-time access to data. This can be extracted from the IN recommendations. A deeper analysis is needed to understand what data types exist in the database. We base the analysis to the IN CS-1 service features and to the GSM frame architecture. A service feature is a building block of an IN service. Since all services are made of service features, analyzing the features gives the data aspects of current and future services.

The analysis of both IN and GSM data is based on the object paradigm. Each analyzed service feature is considered a client for an object-oriented database. The database holds a set of object classes each of which has a set of objects. Each object models a real-world entity, such as a subscriber or a phone line. We adopted this approach already in our early work [Tai94a, Tai94b, RKMT95, TRR95] since object-oriented databases are a natural expansion to relational databases, and the IN architecture is object-oriented.

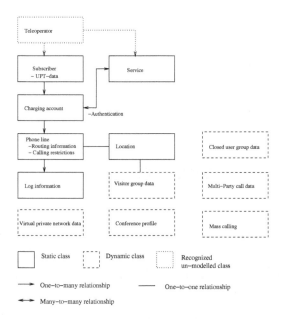

Figure 3.1: IN and GSM object classes.

In the analysis we have identified seven object classes whose objects are time-independent: Teleoperator, Subscriber, Charging account, Service, Phone line, Location, and Log information. We call these static object classes. Objects in these classes may be identified as traditional object-oriented database objects. Next to these we have identified six classes whose objects are time-dependent: Closed user group data, Conference profile, Multi-Party call data, Mass calling, Visitor group data, and Virtual private network data. We call these dynamic object classes. Objects in these classes are dynamic in creation and deletion, and may have timing constraints. Altogether these classes give a basis for analyzing IN/GSM database requirements and solutions. The recognized classes and their relationships are in Figure 3.1.

The IN/GSM static object classes have hierarchical relationships with each other. The only serious exception to the hierarchy is the Service class that is in a one-many relationship to the Teleoperator class, and in a many-many relationship to the Charging account class. The dynamic object classes have a more complex relationships with static objects. We give a detailed analysis of the classes and their relationships in Section 3.1 for IN, and in Section 3.2 for GSM.

3.1 IN data analysis

The IN CS-1 lists 38 service features as listed in ITU-T recommendations [ITU93b] and summarized by Raatikainen [Raa94]. We have divided them into the following categories: Virtual Private Network (VPN)-related features, management-related service features, regular call-related features, service provider-related features, mass calling and televoting-related features, teleconference-related features, and Universal Personal Telecommunications (UPT)-related features. This classification is somewhat artificial since some features may exist in several categories. However, even an artificial structure helps in the data analysis. Since the IN recommendations do not give exact database requirements for the service features, this analysis does not go to a detailed level. It is sufficient for our needs to recognize the most common real-world object classes that must be modeled in the database.

Basically all service features need a calling user profile, a used service profile, and log writing. We will not mention these in the analysis unless the analyzed service feature has a specific treatment for them.

Virtual Private Network service features

A Virtual Private Network (VPN) is a subset of phone lines in a public telecommunications network that behave as if they were a private network. The lines can be accessed from outside the VPN with a common center number. The phone lines in the VPN have short access numbers within the network.

The VPN is one of the major aspects of IN. This can also be seen from the number of VPN service features. They are Abbreviated dialing, Attendant, Authorization code, Closed user group, Follow-me diversion, Off-net access, Off-net calling, and Private numbering plan.

The Abbreviated dialing allows subscribers to shorten a phone number to two or more digits. Next to the VPN, the feature can be used in the regular phone network. The database items needed in this service feature are the abbreviated number, number translation, the possible VPN data, and the subscriber profile. The abbreviated number is first translated to the VPN number, if the operation occurs in the VPN. The VPN number is then translated to a reference to the physical phone line object.

The Attendant allows VPN users to access an attendant position within the VPN for providing VPN service information. The attendants can be accessed by dialing a special access code. The database items needed for this service feature are the VPN data, the access code, and a number translation. The VPN data holds VPN service information. The database must also store access codes for number translation.

The Authorization code allows a VPN user to override calling restrictions of the VPN station from which the call is made. Different sets of calling privileges can be assigned to different authorization codes and a given authorization code can be shared by multiple users. In the database, the VPN data, VPN station data, and the authorization code are needed to fulfill the service feature. The VPN data holds the information of the VPN stations connected to it. Each station has several authorization codes. This operation may be logged to the VPN station data.

The Closed user group (CUG) allows a user to be a member of a set of VPN users who are normally authorized to make and/or receive calls only within the group. A user can belong to more than one CUG. In this way, a CUG can be defined so that certain users are allowed either to make calls outside the CUG, or to receive calls from outside the CUG, or both. The database items needed are the VPN data, CUG data, and user profiles. A VPN data entry has zero or more CUG-data entries, each of which describes one closed user group. Each subscriber has information in the user profile of her currently active closed user groups.

The Follow-me diversion allows a VPN user to change the routing number of her VPN code via a Dual-Tone Multi-Frequency (DTMF) phone. The updated number can be another VPN code or a regular number. The database items needed for this feature are the VPN data and possibly the user profile. The VPN-data holds information about the current routing numbers. The update may be linked to the user profile.

The Off-net access allows a VPN user to access her VPN from any non-VPN station by using a PIN code. Different sets of calling privileges can be assigned to different PINs, and a given PIN can be shared by multiple users. This service feature needs the VPN data and the PIN code. The PIN code is needed to access the VPN. Once the PIN code is verified, the VPN data is loaded and the network is ready for use.

The Off-net calling allows the user to call outside the VPN network. Calls from one VPN to another are also considered off-net. This service feature needs the VPN data and the outgoing number profile. The outgoing line is translated to the actual line number. The VPN-data and number profiles are used in the translation.

The Private numbering plan allows the subscriber to maintain a numbering plan within her private network, which is separate from the public numbering plan. This is actually a management feature, but since it is part of the VPN, it is listed here. The database items needed are the VPN data and the subscriber profile. The subscriber profile holds the needed information. A number translation is needed to and from VPN, among other things.

Management service features

The management service features are designed for subscribers to manage their private data. Each subscriber may manage her data within the allowed limits. We have included authentication service feature to the management service features since it is closely related to management. The IN CS-1 does not specify IN management. Hence, these service features are by no means complete.

The management service features include Authentication and Customer profile management. The former is a general security service feature that is used in most secure services. The latter is a general management service for user information.

The Authentication allows the verification that a user is allowed to exercise certain options in a telephone network. In other words, the feature allows customer authentication via a Personal Identification Number (PIN). This feature needs the user profile and the PIN code. The database must execute the authentication for security reasons. The less identification-specific information is available outside the database management system, the better is the IN security. Both internal and external Authentication requests should be served by the same database management system. The result of the operation is a boolean value describing the validity of the

PIN code. This operation may be logged in the database.

The Customer profile management allows the subscriber to access her service profile, i.e., terminating destinations, announcements to be played, call distribution, and so on. This is the only service feature that allows the subscriber to reach network data. Each subscriber may manage her service profile in the database. Since this is a service feature, all security aspects have been handled in other features. The only database object needed is the subscriber profile.

Regular call service features

The regular call service features are typical to everyday subscriber services. This distinguishes them from service provider service features that are intended for subscribers that offer special services to others (usually with extra charging). A regular call service feature is generally a service feature that concerns two regular phone lines. A service provider service feature operates between several phones and may include several service number transactions and special billing. This division is somewhat artificial since most regular call services are widely used by service providers, and some service provider service features are used in regular services.

The regular call service features are Automatic call back, Call forwarding, Call forwarding on busy/don't answer, Call logging, Call transfer, Call waiting, Consultation calling, Customized ringing, and Terminating call screening.

The Automatic call back allows the called party to automatically call back the calling party of the last call directed to the called party. The database must provide access to the incoming and outgoing phone line information, and the destination number. The called party receives dynamic call data including incoming and outgoing line information. The line information is needed for possible connection restrictions. The destination number included in the dynamic call data may be translated to the actual number.

The Call forwarding allows the user to have her incoming calls addressed to another member, no matter what the called party line status may be. The data items needed are the incoming and outgoing user profiles, and the outgoing number profile. The incoming and outgoing user profiles are needed for security reasons. The user may not forward calls to any destination. Once the operation is accepted, the forwarded number profile includes information of the new number. Circular call forwarding must be recognized.

The Call forwarding on busy/don't answer feature allows the called user to forward particular calls if the called user is busy or does not answer within a specified number of rings. This
is a special case of the Call forwarding service feature.

The Call logging allows a record to be prepared each time that a call is received to a specified
telephone number. The database holds data about the number profile. The calls are logged
according to the number profile to a phone line and subscriber-specific log.

The Call transfer allows a subscriber to place a call on hold and transfer the call to another
location. This service feature may need a number translation.

The Call waiting allows the called party to receive a notification that another party is trying
to reach her number while she is busy talking to another calling party. This feature is stored into
the called number profile.

The Consultation calling allows a subscriber to place a call on hold, in order to initiate a
new call for consultation. This feature does not need database services.

The Customized ringing allows the subscriber to allocate a distinctive ringing to a list of
calling parties. The database holds information about the possible calling parties in the called
line profile.

The Terminating call screening allows the user to screen calls based on the terminating
telephone number dialed. The called line profile holds information about the restricted numbers
or subscriber/line profiles. Alternatively the restriction information may be stored to the called
subscriber profile.

Service provider service features

A service provider in this context is a subscriber who uses the IN services to give special services
to her own customers. The service features needed are related to different types of company
phone services, such as help desks.

We have included Call distribution, Call gaping, Call hold with announcement, Call limiter, Call queueing, Customized recorder announcement, Destinating user prompter, One number, Origin dependent routing, Originating call screening, Originating user prompter, Premium

charging, Reverse charging, Split charging, and Time dependent routing under this category. Naturally all these services can be used by regular subscribers as well.

The Call distribution allows the served user to specify the percentage of calls to be distributed among two or more destinations. Other criteria may also apply to the distribution of calls to each destination. The database must hold the called number profile which includes information and heuristics for the call distribution. The number is translated according to the profile.

The Call gaping allows the service provider to automatically restrict the number of calls to be routed to the subscriber. The database must hold the service provider profile and the service subscriber profile for this feature. The service provider profile holds information about service subscribers. Each subscriber profile of the service provider holds information about call gaping.

The Call hold with announcement allows a subscriber to place a call on hold with options to play music or customized announcements to the held party. The subscriber profile holds information about announcement services. The announcements may be customized depending on the receiver profile.

The Call limiter allows a served user to specify the maximum number of simultaneous calls to a served user's destination. If the destination is busy, the call may be routed to an alternative destination. The user profile and number profile together hold information about the maximum number of simultaneous calls. A number translation may be needed both when the call is forwarded and when the destination is busy.

The Call queueing allows a served user to have calls encountering busy at the scheduled destination to be placed in a queue and connected as soon as a free condition is detected. Upon entering the queue, the caller hears an initial announcement informing the caller that the call will be answered when a line is available. The delayed call is put on a wait queue. When a free condition is detected, the database triggers the incoming switch to allow the connection.

The Customized recorder announcement allows a call to be completed to a terminating announcement instead of a subscriber line. The served user may define different announcements for unsuccessful call completions due to different reasons such as a call arrives outside business hours or all lines are busy. This feature uses mostly the services of a Service Resource Function

(SRF). The called number profile hold information about various announcement types.

The Destinating user prompter service feature enables to prompt the called party with a specific announcement. Such an announcement may ask the called party to enter an extra numbering through a DTMF phone, or a voice instruction that can be used by the service logic to continue to process the call. This service feature uses the services of the SRF. The number profile holds information about various announcement types.

The One number allows a subscriber with two or more terminating lines in any number of locations to have a single telephone number. This allows business to advertise just one telephone number throughout their market area and to maintain their operations in different locations to maximize efficiency. The subscriber can specify which calls are to be terminated on which terminating lines based on the area from which the calls originate. The database must hold the service provider profile and the incoming number profile. The subscriber profile holds information about possible phone groups. When a group receives a call, the group profile holds information about the possible alternatives. A tailored heuristics can then be used to access the correct number.

The Origin dependent routing enables the subscriber to accept or reject a call, and in case of acceptance, to route this call, according to the calling party's geographical location. This service feature allows the served user to specify the destination installations according to the geographical area from which the call originated. The database items needed are the user profile, the number profile, and routing information. The user profile holds information about possible routing plans. The number profile and routing information are used for routing the call to a new phone line.

The Originating call screening allows the served user to bar calls from certain areas based on the District code of the area from which the call originated. The subscriber or receiving phone line profile holds information about banned areas. The incoming phone line profile holds the area code.

The Originating user prompter allows a served user to provide an announcement which will request the caller to enter a digit or series of digits via a DTMF phone or generator. The collected digits will provide additional information that can be used for direct routing or as a security check during call processing. The user profile and called number profile holds information about the wanted digits and their translations. The received digits must be tested against possible typing errors and misuse.

The Premium charging allows for the pay-back of part of the cost of a call to the called party, when he is considered as a value added service provider. The database must hold the caller and called number profiles, charging information and a log. The log is collected during the phone call. The number profiles and charging information together describe the actual costs for both the caller and called phone lines.

The Reverse charging allows the service subscriber to accept to receive calls at its expense and be charged for the entire cost of the call. This can be used to implement freephone services. The incoming call is logged and later updated to the subscriber's charging account. The database holds the subscriber profile and log information.

The Split charging allows for the separation of charges for a specific call, the calling and called party each being charged for one part of the call. From the database point of view this service feature is similar to the Reverse charging. The incoming call is logged and the costs are later updated to both the caller and called charging accounts.

The Time dependent routing enables the subscriber to accept or reject a call and, in case of acceptance, to route this call, according to the time, the day in the week and the date. From the database point of view this service feature is similar to the Origin dependent routing. The user profile holds information about possible routing plans. The number profile and routing information are used for routing the call to a new address.

Mass calling and televoting service features

The mass calling and televoting are special service features that have their own database requirements. The only service feature listed here is Mass calling. It allows processing of a large number of incoming calls generated by broadcast advertising or games.

The Mass calling needs a specialized database that is tailored for a huge number of short simultaneous write requests. Each request must be stored very fast to allow maximum request throughput. The incoming calls are stored as (called number,vote) -pairs and analyzed online.

The called numbers may be restricted for a predefined number of votes (usually one). The data items can be written sequentially. A more structured storing method may be done after the mass calling is over.

This service feature can easily overload even a very fast DBMS. Due to this it should not be allowed to execute in a regular IN database. A DBMS that is tailored for fast writes is a working alternative.

Teleconference service features

A teleconference in this context is a resource that can be accessed by two or more subscribers. The conference resource is reserved to a certain time at which the members may join it.

The IN lists two service features that are related to teleconferencing: Meet-me conference, and Multiway calling.

The Meet-me conference allows the user to reserve a conference resource for making a multi-party call, indicating the date, time, and conference duration. At the specified date and time, each participant in the conference has to dial a designated number which has been assigned to the reserved conference resource, in order to have access to that resource, and therefore, the conference. The database must hold the conference resource, multi-party call data, and calling number profiles. The conference resource and multi-party call data together hold the conference information. Number profiles are needed for accessing the resource.

The Multiway calling allows the user to establish multiple, simultaneous telephone calls with other parties. This service feature does not directly need any database services. The simultaneous telephone calls may trigger any IN service features.

Universal Personal Telecommunications service features

The Universal Personal Telecommunications is intended for giving a single phone number that can be used everywhere. The number is automatically connected to different phone lines depending on the date and time. The service is not fully defined in CS-1, and it holds only one feature: Personal numbering.

The Personal numbering supports a Universal Personal Telecommunications (UPT) number that uniquely identifies each UPT user and is used by the caller to reach that UPT user. A UPT user may have more than one UPT number for different applications. However, a UPT user will have only one UPT number per charging account. The database items needed are the UPT subscriber profile, UPT data, and a charging account. The UPT data holds UPT-related information about the account. The user profile holds subscriber-specific information. A charging account is needed for each UPT number.

IN data summary

Although we have analyzed nearly forty service features, the number of classes found in them is relatively small. We will first list the static classes and their properties.

A *teleoperator* offers services to subscribers and also to other teleoperators. Hence the Teleoperator class is the core of the IN data hierarchy. Although it is not explicitly listed in the features, it is the class on which an IN operator may access other networks. A Teleoperator has zero or more services to offer, and zero or more subscribers as customers. However, we are not that much interested in the services and subscribers of all teleoperators. Rather we want to have an entry upon which to request more information. Due to this the Teleoperator class is recognized but not modeled.

All service features are initiated by a *subscriber* which gives us the class Subscriber. The Subscriber class holds information about each customer of the teleoperator. A special sub-scriber subclass is a Service provider who has extra privileges (with extra costs). The Universal Personal Telecommunications data belongs to the Subscriber class since it is unique to a single subscriber.

A subscriber may have several *charging accounts* each of which has a set of IN services. A charging account has a set of IN services and a set of charged lines. This class may be combined into the Subscriber class but that would cause unwanted data replication.

Each account of a subscriber has a set of active *services* offered by her teleoperator. A Service class holds information that is relevant to a single service. It has links to the charging accounts that have ordered this service.

Each charging account has zero or more *phone lines* that can be accessed directly via a phone number or indirectly via a number translation. A phone line belongs to exactly one charging account and has the services that are active in the charging account. The line information class has routing information and call restrictions.

Whenever a phone line is accessed, a *log information* entry is created by the accessing transaction.The log data is written in blocks and read when necessary directly or via the phone line object.

The previous structure is fairly hierarchical. A teleoperator has zero or more subscribers. A subscriber has zero or more charging accounts. A charging account has zero or more phone lines, and a log is created for every access to a phone line. On the other hand, a teleoperator has zero or more services and a service may be subscribed by zero or more charging accounts. Hence, we get two access hierarchies: one from a subscriber and one from a service. Sometimes also a direct access to a phone line and perhaps to log information is necessary. But altogether the major IN classes behave in a hierarchical manner.

Next to the regular classes that are fairly well organized and static, a set of dynamic classes exist. These classes break the hierarchy by referencing information at several levels. The classes are VPN-data, Mass calling data, Conference resource, Multi-party call data, and CUG-data..

The *VPN-data* class holds information about Virtual private networks. It is linked to Phone lines, Subscribers, and allowed Services. This class is usually fairly static but dynamic VPN-cases may also exist.

The *Mass calling data* class is needed only for mass calling and televoting. It stores blocks of (called number,vote)-pairs. It is not directly related to any other classes although it can access the called number data entry and via that the Subscriber data. Usually this is not necessary since in televoting the voting privacy is as important as in regular voting.

The *Conference resource* class holds information that is needed to hold a teleconference. It has links to conference Subscribers and Phone lines. It is relatively dynamic in nature since its life span is equal to the length of the corresponding conference.

The *Multi-party call data* holds information about the members of a call conference. It is very dynamic in nature since in principle the participants of the conference are allowed to change during the conference.

The *CUG-data* (closed user group data) is very dynamic in nature. It holds information about VPN members that belong to a specific closed user group. The group may send and receive calls only from other members. The object exists only for the life span of the closed user group itself which can be very short.

These dynamic class types do not behave well in the earlier static hierarchy. They either access objects at random levels of the hierarchy (such as VPN-data) or do not access any objects (such as Mass calling data). Their dynamic nature makes them behave more like special views to the static data.

3.2   GSM data analysis

Unlike IN data analysis, the GSM data analysis cannot be based on services. The GSM network is mainly designed for fast and reliable access for mobile stations. While a GSM operator offers special services, the services are built on top of the GSM network instead of inside it. Due to this, we must analyze the GSM data needs from the mobility point of view. It is most probable that any mobility aspects in future IN will be based on the tested and working aspects of GSM networks. As such, the GSM gives an excellent framework upon which to expand current IN data requirements.

The three subsystems in GSM, BSS, NSS, and OSS, have different database needs starting from no requirements and ending in strong real-time requirements.

The BSS does not have any database requirements. If the BSS architecture uses databases, the databases are embedded into BSS elements and are vendor-specific. The BSS architecture does not affect the requirements.

The NSS manages the communications between the GSM users and other telecommunications network users. The NSS has four major database management systems that are listed in the GSM standards: HLR, VLR, AC, and EIR.

The HLR is the main database management system in a GSM network. It is an active component of the system and controls call and service related information. The HLR must offer real-time database services. Its main functions include such time-critical tasks as finding sub-scriber information, informing the current location of a subscriber, billing and service control. Usually the architecture of HLR is not distributed. A HLR can handle hundreds of thousands subscribers. If more capacity is needed, several HLRs can be executed in parallel. This can be achieved since subscriber data is hierarchical. The VLR has detailed data on location and service data regarding MSs that are in its coverage area for routing. This data is then used for incoming and outgoing calls. The VLR services include location updates, temporary mobile station identity allocation for location confidentiality, and storage of mobile station roaming number. The VLR has a much smaller role than HLR. It is responsible for maintaining ex-act location information of subscribers that are in the area of its Mobile Switching Centers. It communicates with the HLRs when calls are routed or charging information must be retrieved. Hence it can be considered as a specialized database node in a distributed real-time database. It has real-time and distribution requirements. However, the distribution requirements are not strict. Updates occur only in one VLR, and the only distribution is into known HLRs. Since the information location is always known and a single HLR, distribution basically degenerates into message exchange. The AC is used to verify the authorization for required GSM services. As such it supports the same service as the IN authentication service feature. The AC needs to serve real-time requests, but it does not need distribution. It is often connected to the HLR, or a single database management system offers both HLR and AC services. Finally, the EIR maintains information about MSs. It can be distributed since several EIRs may co-operate.

The most important database managers in GSM are HLR and VLR. The HLR always holds static information about each subscriber. When a subscriber roams, the roaming information is held on the corresponding VLR. Hence, a HLR has connections to several VLRs on different networks and the connections are dynamic in nature. Together the HLR and VLR have the following interactions:

- Roaming from one VLR area to another. This changes the subscriber location information in HLR and changes the VLR that holds location information.

- Routing a call from a subscriber to a mobile station and receiving a call from a subscriber to a mobile station. Both of these need interaction between the HLR and the current VLR. The HLR is responsible of the routing while the VLR gives enough location information to make the routing possible.

- HLR and VLR failure and relocation procedures. These are needed when a system or network failure occurs.

- HLR subscriber information update. This occurs when a subscriber is added to or deleted from a HLR, or subscriber information is changed. The changes are reflected to the corresponding VLR, for instance when a subscriber is deleted.

- VLR information update. This occurs when a subscriber roams from one VLR coverage area to another. The change is informed to the HLR.

The HLR architecture is best described as a parallel database architecture where several database nodes are connected together with a fast bus to form a logically uniform database management system. It exchanges information and messages with a set of VLRs, some of which are in external networks. The HLR has distributed requests but they degenerate into message passing between the HLR and one or more VLRs. Data is not replicated and data location information is well known. All real-time reads occur only in a single node, since the needed information is subscriber information which is hierarchical.

When a mobile equipment holding the mobile phone line is within the network of the subscriber's teleoperator, location information is updated continuously to mobile phone line data. During this time the network has knowledge of the mobile phone line location. If the connection is lost (either the mobile equipment was closed or the equipment is no longer within the reach of the network), the network has knowledge of the last location.

Sometimes a subscriber may roam to an area that is no longer part of the subscriber's teleoperator network. In such a case, when an external network detects the mobile phone line, it asks the mobile phone line for its identity. The identity holds information about the subscriber's home network.

Each teleoperator has information about other teleoperators. If an external network operator and the home network operator have a roaming contract, the external network contacts the subscriber's home network and informs it that this mobile phone line is in its area, and it informs the mobile phone line that external network services are available. After that the identity of the visiting mobile phone line, location information, and link to the home information, is stored to a VLR.

Once a subscriber of a visiting mobile phone line wants to use any of its services, including initiating a phone call, the service requests are routed via the home network. This occurs as long as the mobile equipment is within the network of the external teleoperator.

Later, when the mobile equipment leaves the network, the external VLR informs the home network that the subscriber has left their network. The mobile equipment may start another session in a new network, and the previous recognition procedure happens again. Thus, during the time between the two networks, the home network has only knowledge of the last known location of the mobile phone line.

The previous paragraphs describe the GSM visible database services, their distribution and real-time aspects. While the GSM network includes several special services, such as Short Message Services (SMS) and WAP services, they are built on top of the GSM network and are not very interesting to us. They use special communication protocols and service databases. The former is a network issue, the latter can be included to service data classes.

The OSS manages the GSM network. It includes subscriber management, system mainte-nance, and system engineering and operations. The OSS can have several database management systems, although the GSM standard does not specify them. The management databases need not to be real-time databases. Although requests must be served rapidly, it is enough that they are served with high throughput databases rather than real-time databases with deadlines.

The subscriber management service includes subscription, billing and accounting manage-ment. The subscription management is used when subscribers' data is changed or a new sub-scriber is added to the network. The billing and accounting management are used when a subscriber is billed of the network use. While these services can be very complex in nature, they do not add anything new to the GSM object class analysis.

System maintenance includes failure detection and repair of GSM components. As such it has a database of all GSM network elements. This is closely related to network element management in IN or TMN. System engineering and operation is closely related to the system maintenance. It includes tasks to design and optimize the GSM network architecture. As such it needs similar database services as the network element management in system maintenance.

The GSM data items and IN data items have a direct correspondence with each other. The GSM subscriber entity is related to IN subscriber, phone line, and charging account entities. The GSM service entity is equal to an IN service entity. The GSM location entity is an expansion to IN entities but it is easy to embed in the IN phone line class. The GSM authentication entity can be embedded in IN entities since the same service feature is defined in IN. The only really new entity is the Equipment entity. It is more of a management entity and does not need special database treatment. In fact it may be the best to implement as a distinct traditional database management system.

The data analysis also shows the benefits of distribution. Since the combined IN/GSM class structure is mostly hierarchical, it is possible to fragment data into several nodes and still have mostly local transactions. This is a useful property because distributed transactions always cause more overhead than local transactions. With such a hierarchy it is possible to truly benefit from distribution instead of losing processing power to the overhead of distributed transactions.

Chapter 4

Yalin architecture

The IN/GSM data analysis in the previous chapter gives us a basis on which to design our architecture. The identified classes have a nice hierarchical relationship, most IN/GSM services are time-driven, and some information has a temporal nature. Due to these facts a good candidate for an IN/GSM database management system is a RT-DBMS.

A chosen RT-DBMS architecture must support both IN/GSM database requirements and distribution [TR00]. A large-scale real-time distributed database management system can of course fulfill all real-time and distribution requirements. Unfortunately an implementation of such a system is not a reasonable approach since this would also add unwanted functionality that affects general database performance [TS97b, TS96, TS97a]. Requiring both real-time and distribution in the IN/GSM database management system causes at least three issues.

First, the transaction deadlines in an IN database are often very short. Even the geographical latency of a distributed database can be too much for the shortest deadlines. Although it may be possible to store data near the requesting client by using replication, some compromises must be made between distribution and replication. A heavily replicated distributed database solves read problems but causes update and volume problems. A non-replicated distributed database needs only a single update and optimizes the data storage volume, but then reads may suffer from distribution latencies.

Second, the data network reliability is an issue. If the IN/GSM database is to be fault-tolerant, the connections between distributed nodes must also be fault-tolerant. This can be achieved to a certain degree with replicated connections between network elements, but such a solution soon becomes very expensive. It has been shown that in a distributed network that is not fault-tolerant it is not possible to create a non-blocking atomic commit protocol [BHG87]. This affects short deadlines. Also during the recovery time the system cannot fulfill its deadlines. Although SS7 is reasonably reliable, network failures still occur.

Third, a real-time distributed database allows any kind of semantic connections between data items. The IN/GSM data structure is hierarchical with few non-hierarchical relationships. This information can be used to optimize the IN/GSM architecture and data locations. Building a general RT-DBMS is a much more challenging task than building a tailored RT-DBMS that benefits from known semantic relationships between object classes.

Generally we want an architecture where we can minimize distribution of real-time transactions. Every distributed transaction causes overhead and unpredictability to the system. We can tolerate it to some extent when the system is not heavily loaded and when deadlines are relatively long. However, these requirements are not met in regular IN/GSM transactions. The load of the system can be very high at rush hours and deadlines are short.

Starting from the listed issues above, we have designed a reference architecture Yalin (Yes, A Lovely IN database architecture). This architecture is based on the IN/GSM data model and as such it is a very specialized one. We can use the analysis of the IN/GSM data hierarchy to justify our architecture.

The IN/GSM data structure is hierarchical. It is also easy to translate into physical entities. Just like in regular distributed databases where data is divided into a set of fragments, we can divide the IN/GSM data into several fragments. We call this the fragmentation of the Yalin database (Figure 4.1).

The fragmentation of the Yalin database is specialized since the IN/GSM data is also specialized. It consists of a root fragment, a set of regular fragments, a set of VLR fragments, and a set of Mass Calling fragments. The only non-fragmented class is the service class. Services are replicated to regular and root fragments.

The root fragment has teleoperator-related information. This information includes both services and subscribers, and also links to other teleoperators. The root of the IN/GSM data hier-

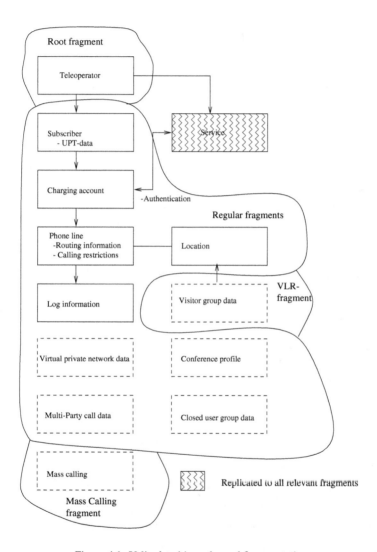

Figure 4.1: Yalin data hierarchy and fragmentation.

archy resides in the root fragment. The other fragments have links to the root fragment.

The regular fragments have a subscriber-based view to the IN/GSM data hierarchy. The actual classes are divided into one or more fragments. Hence, each fragment holds part of the IN/GSM data and most of the IN/GSM object classes. Each regular fragment has a link to the root fragment which serves as the root of the tree. The regular fragments can be further divided into new fragments. Each new fragment is an equal subtree whose root is the root fragment. This is a logical fragmentation policy in the IN/GSM database because the data itself is accessed hierarchically. Any other fragmentation policy would cause unnecessary distributed transactions for each connection.

Next to the regular fragments, two special fragment sets are defined: VLR fragments and Mass calling fragments. The VLR fragments hold data that is relevant to VLRs in the GSM physical model. These fragments are separated from the regular fragments since their data is of subscribers from external networks. The Mass calling fragments are only for mass calling and televoting data. This data, while clearly part of the IN data hierarchy, is specialized enough to be stored in separate fragments. Data on the Mass calling fragments is linked to the subscriber and service data when necessary. The data volume of mass calling and televoting fragments may grow extremely fast.

The fragmentation of the Yalin database is simple to translate into a database management system architecture. Each fragment belongs to one node in Yalin, and a node in Yalin has one fragment. Together the nodes form a distributed database management system.

The nodes of the fragmented Yalin database have clearly various types of data volume and access requirements. The size information of Yalin fragments affects architectural decisions of the Yalin database.

The Root fragment has information of teleoperators and services. It is a tiny fragment which in principle can physically reside in one of the regular fragments. It is usually accessed from external networks since local requests are easier to handle directly in the regular fragments.

The VLR fragment holds information about mobile visitors in our network. This specialized information is relatively large in size. Fortunately this is a rather separated entity in the architecture since VLR data is seldom needed in the regular IN/GSM fragments. Thus, this fragment is usually accessed directly when VLR services are needed.

The regular fragments hold normal classes and dynamic groups. Most of the classes are small or medium in size. An exception here is the log information class which can grow rapidly to very high volumes. The regular fragments may be accessed directly or via the root fragment.

Finally, the mass calling fragment has very specialized access requirements. During the mass calling phase transactions only write to the fragment. Once data is analyzed, transactions only read from the fragment. This fragment is usually accessed directly.

Since we have both data volume and data access requirements that are specific to certain types of fragments, it is natural to have a specialized architecture for each of the fragments. The small fragments are easy to design within a single database node, but the larger ones need a parallel database architecture that we call a *cluster*. It is a parallel database that consists of a set of shared-nothing parallel nodes. It behaves like a single database node in a distributed database. This approach allows a cluster to be scalable. Large clusters have several simple nodes. Small clusters may need only one node.

The Yalin architecture can be seen in Figure 4.2. It consists of three abstraction levels: database, cluster, and node level. A local request may arrive to any of the levels. Requests from external networks are restricted to the database level.

The database level abstraction gives a standardized view of the Yalin database. In the current IN recommendations the standard view is based on the subset of the X.500 directory protocol [CCI89] which is also the basis of the database level abstraction [ITU96]. The database abstraction view of the Yalin database offers the services of a X.500 Directory System Agent (DSA). This abstraction is defined mostly for external requests from untrusted sources. All trusted sources should use the lower abstraction levels for faster access.

The X.500 directory protocol is needed for compatibility reasons both for requests from external networks and for applications that want a standard application interface to Yalin. The database does not have to use X.500 internally, and even local IN service requests may be served directly without the X.500. A private interface is the primary entry to the database for trusted applications. A verification system, such as a PIN code, may be necessary for some applications.

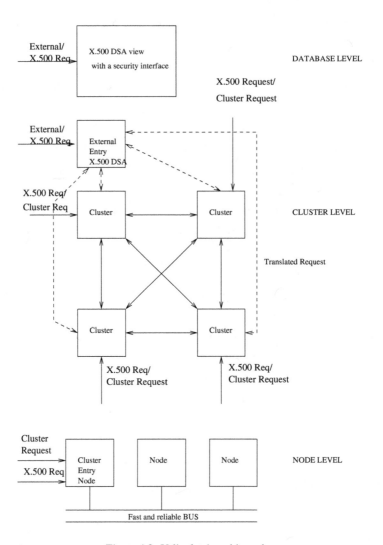

Figure 4.2: Yalin database hierarchy.

The cluster level abstraction gives a distributed view to the Yalin database. It is best suited for transactions that must access data in several clusters. This abstraction defines an entry node for X.500 requests and a set of specialized clusters. Each cluster corresponds to one fragment in the Yalin data fragmentation. The abstraction defines both an X.500 based access and an internal optimized access to clusters. This way also standard IN-applications may have cluster level access.

The node level abstraction is the lowest abstraction level. At this level the internal structure of a cluster is visible. This is also the fastest access level. Each cluster consists of a cluster entry node for database and cluster level requests. The entry node is responsible for forwarding the requests to appropriate nodes. Each node in a cluster can also be accessed directly. The nodes in a cluster are connected with a fast and reliable bus. This architecture is basically a parallel shared-nothing database architecture with an entry node.

4.1 Database entry node

The Yalin Database entry node implements the database level abstraction of the Yalin database. It offers an interface for X.500 based requests that are either from untrusted clients or are from local clients that want to use the standardized IN interface to access data. The entry node is the only way that external clients may use to access the database. For local clients, the cluster level abstraction offers an X.500 interface on each cluster.

Next to the interface services, the database entry node interprets the received X.500 requests to internal transaction requests, offers root DSA services to other DSAs in clusters, co-operates with other DSAs when necessary, and co-operates with other X.500 root DSAs in external networks. It maintains X.500 level directory information about data locations and information about external network entry nodes and teleoperators.

The most important task that the database entry node fulfills is to offer a known entry for external clients to the network. As such it is mostly an X.500 root agent that co-operates with other X.500 root agents of external networks. When we have a well-known entry point into the database, the rest of the architecture can be hidden from external clients.

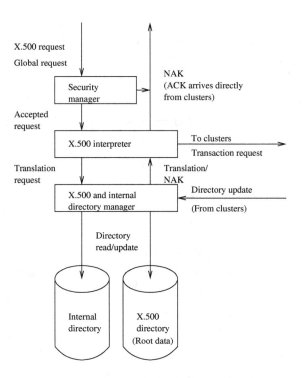

Figure 4.3: Database entry node.

The architecture of the entry node consists of a Security manager, an X.500 DSA interpreter, and an X.500 and internal data directory manager (Figure 4.3). The security manager accepts requests from applications. It is responsible for checking the access rights of a requesting client. Both external and local clients are checked when a request arrives. When the request is accepted, it is forwarded to the X.500 DSA interpreter. The X.500 DSA interpreter is responsible for interpreting an X.500 request to an internal request that can be sent to clusters. In order to do this the interpreter consults the X.500 and internal directory manager that has access to interpretation data from the X.500 data tree to the Yalin data tree. Once an interpretation is complete, the interpreted request is sent to one of the clusters. Once the request is served, the result is sent directly to the requesting application. The X.500 and internal directory manager manages both an X.500 specific data directory and a Yalin specific data directory. Both these directories are plain directories. Actual data resides in clusters. When directory data needs an update, an update request is sent to the entry node.

The Yalin architecture allows several database entry nodes. This is probably not necessary since the database entry node serves mostly requests to and from external networks. Since the tasks of the entry node are relatively simple, with proper hardware a single node should easily be able to handle all external network traffic. If this is not the case, adding fully replicated database entry nodes can be used to ease bottlenecks in the X.500 data model.

4.2   Clusters

A cluster in Yalin behaves externally like a node in a distributed database. Each cluster has a similar interface regardless of the data fragment that it supports. The interface supports both X.500 requests and internal Yalin requests. All X.500 requests are interpreted to direct requests, so the clients should use the internal interface to Yalin whenever possible.

The cluster architecture consists of a cluster entry node and a set of database nodes. The entry node co-operates with other clusters to create and maintain distributed transactions. Actual data processing occurs in the database nodes. The architecture of a cluster can be seen in Figure 4.4. It consists of elements that are in the cluster entry node: an X.500 interpreter, a Cluster

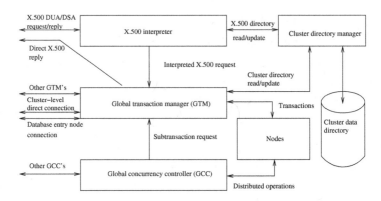

Figure 4.4: Cluster entry node.

directory manager, a Global transaction manager (GTM), and a Global concurrency controller (GCC). The nodes element which is also present in Figure 4.4, represents the database nodes in the parallel database architecture.

The cluster entry node behaves like an X.500 DSA. The X.500 requests are interpreted to Yalin internal transaction language in the X.500 interpreter and then forwarded to the GTM. The X.500 interpreter consults the Cluster directory manager which holds directory information about cluster data. The interpretation is straightforward when the clusters follow the IN/GSM data hierarchy. If the hierarchy is broken, the interpretation may create heavily distributed transactions.

Sometimes the X.500 interpreter receives X.500 directory update requests from other DSAs. They are interpreted to the Yalin internal transaction language and forwarded to the Cluster directory manager. It updates directory information accordingly.

When an X.500 request is interpreted, or when a direct Yalin request is received, it arrives to the Global transaction manager. The GTM creates a new transaction from the request and then consults the Cluster-directory manager to find out which node is best suited for this transaction. The transaction is then sent to the appropriate node which processes it.

The nodes execute the created transaction. During the execution phase the transaction may need to reference data in other clusters. The Global concurrency controller receives a read/write request which it will forward to the appropriate subtransaction in another cluster. If the request is the first one to this cluster, the GCC sends a subtransaction request to the GTM. The GTM forwards the request to the appropriate cluster. Sometimes the GTM can already decide that the created transaction needs to access several clusters. In such a case it sends the subtransaction requests to the appropriate clusters directly. Naturally both a set of known subtransactions and requests for new subtransactions may exist in the same distributed transaction. First a set of subtransactions is created to other clusters and later the GCC sends a subtransaction request to the GTM.

When a transaction is ready to commit, it participates in a global commit procedure with other subtransactions. If the transaction was responsible for the creation of subtransactions, it becomes the leader in the atomic commit. Otherwise it is a subtransaction and it participates as a follower.

The GTM consults the GCC to check transaction conflicts. If the transaction conflicts with other transactions, and the conflict is illegal to the chosen concurrency control policy, either the committing transaction is aborted or the conflicting transactions are aborted. The chosen approach depends on the global concurrency control policy. Since conflicts are checked only at commit times, deadlocks cannot occur. This behavior is typical to all optimistic concurrency control policies.

Since clusters support global concurrency control, they also support location transparency. The result of a deterministic request is the same no matter which cluster receives it. However, for efficiency reasons the clusters also support location forwarding. That is, when it is obvious that a request can be served better in some other cluster, the receiving cluster forwards this information to the requesting client. The client can either accept the suggested cluster and re-send the query there, or it can save this information and execute this request as before.

A created transaction may change the tree structure of the cluster. For instance, a branch of the cluster tree may be moved from this cluster to another. In such cases the GTM sends the update information to the root node and to the GTMs in other clusters. This information is needed to maintain the global X.500 based view of the database. However, the cluster-specific moves are not forwarded to the root node. The root node does not need to know about data locations inside a cluster.

While the clusters in Yalin are specialized, the cluster level architecture is the same on all the cluster types. The difference between the clusters is in their use. Regular clusters are for regular requests. The mass calling and televoting clusters may not be always accessible, but when they are needed, the volume of the requests is so high and specialized that the requests should not be allowed to affect transaction throughput in regular clusters. Finally, the VLR clusters are not part of the IN/GSM data tree. Rather, they hold information about visitors from external networks.

4.3   Nodes

The node architecture in Yalin is the core element in the database. Where the other elements are mostly transaction interpreters and organizers, the nodes handle database management system operations.

Each node may receive two types of transaction requests. First, cluster requests arrive from the cluster entry node. These requests are originally from the X.500 root node, from cluster-level clients, or from intra-cluster transactions. Second, a node may receive local requests. These requests are mostly for maintenance staff although in principle it is possible that a client wants to access a cluster directly and thus bypass the cluster entry node.

The nodes in a cluster behave like a set of nodes in a parallel shared-nothing database. This solution allows expandable database nodes and transaction distribution between nodes inside the cluster. It also simplifies expansions both at the cluster level and at the node level, since adding a new distributed database cluster expands Yalin at the cluster level, and adding a new parallel database node to a cluster expands Yalin at the node level. The former expansion corresponds to a situation where the teleoperator opens a new branch to a new location. The latter expansion corresponds to a situation where the teleoperator has to increment the database capacity due to a customer volume increase in a local area.

At minimum, a cluster has a single node which is embedded into the cluster entry node. This architecture can be seen in Figure 4.4. The number of nodes in a cluster is limited only by

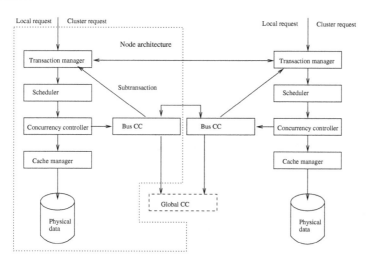

Figure 4.5: Regular node architecture.

the hardware. The cluster level managers and transactions see only a single node architecture.

The node architecture follows general database architecture structures, including a Transaction manager (TM), a Scheduler, a Concurrency controller (CC), a Bus concurrency controller (BCC), and a Cache manager (CM). It is not meant to be a complete database management system itself. Instead, it is a framework upon which to build the remaining elements. The node architecture can be seen in Figure 4.5.

The TM accepts local and cluster level transaction and subtransaction requests. They all get similar treatment in the manager. A transaction request creates a new transaction. A subtransaction request creates a new parallel database subtransaction that has a leader in some other parallel database node in the cluster.

When a transaction or a subtransaction has finished, the TM manages bus-level atomic commit procedures and local commits. The result of the commit or abort is then forwarded to the cluster entry node. If the transaction is a participant of a distributed transaction, the final result of the commit status depends on the cluster level vote results. Thus, a transaction may be aborted at the cluster level even if it can commit at the node level. If the transaction is not a participant in a distributed transaction, the cluster level transaction manager accepts the commit immediately.

When the TM has created a transaction, it sends the created transaction to the Scheduler. The Scheduler manages the transactions that are either executing, waiting for I/O, or ready to execute. Each transaction receives the CPU according to the chosen scheduling policy.

A transaction may have three types of operations: read operations for reading data from the database, write operations for writing data to the database, and execute operations for calculating intermediate and final results from read data. These three operations are sufficient to model scheduling and concurrency control in Yalin.

The read and write operations are forwarded to the CC. The CC keeps track of data items and transaction operations. It consults the CM when data has to be moved to and from the disk. If the requested data item does not exist in the local node database, the CC consults the BCC to see if the item exists in some other node. If this is the case, a new subtransaction is created and the transaction becomes distributed within the bus nodes. The BCC may also consult the GCC in the cluster entry node if the data item does not exist in any of the nodes. In such a case the transaction becomes an intra-cluster distributed transaction. We will cover node level concurrency control in Section 4.5. Finally, the CM manages physical data in the database. It has only two operations fetch and flush. The fetch operation reads a physical data item from the disk to the main memory. The flush operation writes a physical data item from the main memory to the disk. The CM is the only manager that can access physical data.

The standard node architecture is stripped and follows general transaction management principles. It is better to have a stripped but nevertheless a working model that we can analyze than a fully equipped model where interesting analysis is lost into a number of uninteresting details. Due to this we have omitted such managers as recovery managers, query interpreters and optimizers.

4.4   Database transactions

In the IN/GSM data analysis in Chapter 3 we showed that the IN/GSM data hierarchy includes the following classes: Teleoperator, Service, Subscriber, Charging account, Phone line, Log information, Mass calling data, Virtual Private Network (VPN) data, Conference resource, Multiparty call data, Closed user Group (CUG) data, Location information, Visitor group data, and Equipment identity register (EIR) data. This is a simplified model of a telecommunications database classes but it still gives us a basis to define major transactions for Yalin.

The transaction types of Yalin can be identified from the IN/GSM service feature analysis and the Yalin database schema. In the analysis we have identified the following transaction types: IN service activation transactions, number translation transactions, mobile equipment location update transactions, other mobile related transactions, query transactions for management, line profile transactions, call triggered transactions, charging transactions, dynamic group related transactions, management update transactions, and EIR-specific transactions.

Regular service related transactions are used when a subscriber wants to activate an IN-service. They are real-time transactions and common in the system. They need read-access to Service, Subscriber, and Charging account classes.

Next to the regular service related transactions, we have transactions that handle all kinds of number, charging and phone line translations. These transactions are real-time and common in the system. They need read access to Service, Subscriber, Charging account, Phone line and Log information classes.

A typical transaction is a location update transaction. It is used every time a mobile equipment signals its current location to the network. The transactions are real-time and common in the system. They need write access to the Location information class.

Communication transactions between a visitor's HLR and a visiting VLR are mobile related transactions. They are relatively common in the system and real-time. They need write access to Visitor group data, Location information, and Phone line.

The management staff needs query transactions. The transactions need read access to all classes. They may be long and distributed. The query transactions are not real-time and not very common in the system.

Whenever a customer opens a phone line, a line profile transaction is triggered. It is used for receiving information about the services and phone lines of the customer. The transaction needs read access to Service, Subscriber, Charging account, and Phone line classes.

A call triggers at least three transactions. First, when a call is initiated a transaction is created to gather IN and routing information, and to create a call log entry for the call. Second, when a call is answered a transaction is created to gather IN information and to update the call log entry for this call. Third, when a call is finished a transaction is needed to update call log information and decide charging. The Init call transaction needs read access to Location information, Service, Subscriber, Charging account, and Phone line classes; and write access to the Log information class. The answer call transaction needs similar read and write access, except that the Location information is no longer needed since the call is already connected. Finally, the finished call needs read access to Service and Phone line classes, and write access to charging account and Log information classes.

At specific times a subscriber receives a bill of her charging accounts. The operation triggers a charge transaction that reads and updates Charging accounts and perhaps needs read access to services and subscribers as well. Thus we get a transaction that needs read access to Service and Subscriber classes, and write access to the Charging account class. In principle the transaction may also need read access to Log information, for instance when a detailed bill is to be created.

A special set of transaction types is needed to manage dynamic groups: Mass calling data, Virtual private network, Conference resource, Multi party call data, and Closed user group. Mass calling data needs two transactions: one to write a new entry and one to analyze the entries. The former needs write access to the MCD dynamic group. The latter needs read access to the MCD group. Virtual private network needs number translations and call connecting. These types of transactions have been defined earlier. A management transaction of the VPN is also needed. It needs write access to the VPN dynamic group. Conference resource needs a transaction to manage an incoming conference, and another transaction to create a new conference. Multi-party call data is needed together with conference data. The transactions to manage it are the same types of transactions that are needed in the conference resource. A closed user group

Type	T	S	Su	CA	PL	LI	VG	Lc	MC	VPN	CR	MP	CUG	EIR
Activate		R	R	R	R									
Translate		R	R	R	R	R								
Loc upd								W						
Mobile	R						W	W						
Query	R	R	R	R	R	R	R	R	R	R	R	R	R	R
Profile		R	R	R	R									
Init call	R	R	R	R	R	W		R		R			R	
Answer call		R	R	R	R	W				R			R	
Finish call		R		W	R	W				R			R	
Charge		R	W			R								
Mass write									W					
Mass read									R					
VPN mgmt										W			W	
Conf mgmt											W	W		
Create conf											W	W		
Add member												W		
CUG mgmt													W	
Top mgmt	W													
Serv mgmt		W												
Subs mgmt			W	W	W									
Acc mgmt			R	W	W									
Phone mgmt			R	R	W									
EIR mgmt														W

T	Teleoperator	S:	Service
Su:	Subscriber	CA:	Charging account
PL:	Phone line	LI:	Log information
VG:	Visitor group	Lc:	Location
MC:	Mass calling	VPN:	Virtual private network
CR:	Conference	MP:	Multi-party call
CUG:	Closed user group	EIR:	Equipment identity register

Table 4.1: Transaction types and telecommunication classes.

class is needed in CUG-services of a VPN. The transactions needed to manage the class are the VPN management transactions and a transaction that has write access to CUG data.

A set of management transaction types is needed to manage such classes that are not directly managed by other types of transactions. These include Teleoperator, Service, Subscriber, Charging account, and Phone line. Each of these classes need a write access transaction for class creation and management. Finally, a transaction type is needed to manage the EIR class. Thus such transactions need write access to the EIR class. All these management transactions are relatively rare in the system and are not real-time.

The summary of the transactions and their read/write access to the Yalin schema is in table 4.1. The table clearly shows that most conflicts in the system occur in charging account, phone line, and log information classes. Other object classes are almost free of conflicts. However, even in these busy classes the conflicts are relatively rare due to the large number of objects

in the classes. Along with the real-time requirements, the transaction analysis justifies optimistic concurrency control methods in Yalin. If conflicts were more common, a pessimistic
concurrency control with priorities would be a reasonable alternative.

4.5   Database concurrency control

Since we have short transactions with tight timing constraints, it is important to minimize the
resources needed to manage transactions. This includes resources needed for concurrency control. One very important area in this is global concurrency control.

In principle we can use any concurrency control policy in a distributed database. All we have
to do is to ensure that transactions behave well both locally and globally. This often leads to a
two-level concurrency control management: local concurrency control and global concurrency
control.

Local concurrency control management maintains the needed transaction isolation level
within nodes. It does not differ from concurrency control in non-distributed databases. Global
concurrency control management maintains the same isolation level between nodes.

Usually local concurrency control management cannot ensure that transactions do not interfere with each other globally. Due to this unfortunate feature distributed databases need to
take extra care in order to maintain the desired isolation level both globally and locally. It can
be done for instance by sharing serialization graphs between nodes and always checking if the
combined graph has a cycle. This can become very expensive.

In Yalin we wish to avoid global concurrency control as much as possible. From the database
and transaction analysis earlier we can see that the number of distributed transactions is small
compared to the total number of transactions. Most of the time distributed transactions do
not interfere with each other. Yet every distributed operation may cause a cycle in the global
serialization graph and thus violate the global isolation level of Yalin.

Fortunately it is possible to avoid global concurrency control management completely when
each node in a distributed database follows a strict concurrency control policy. In such a case
local concurrency control management is sufficient at the global level as well. We will prove

this in the next paragraphs.

Theorem 4.1 *A concurrency control policy is strict if for every history H that is allowed by the policy the following holds: whenever $w_j[x] < o_i[x](i \neq j)$, either $a_j < o_i[x]$ or $c_j < o_i[x]$ [BHG87]. Let us have a distributed database with nodes $N_1 \ldots N_n$. If on every node N_i the local concurrency control policy is strict then transactions also serialize globally.*

Proof: The strict concurrency control policy ensures that transactions are only allowed to reference data that has been written by a committed transaction. This ensures that all local serialization graphs $SG_1(H) \ldots SG_n(H)$ are acyclic [BHG87].

Let us have two distributed transactions Tr_1 and Tr_2. If both transactions are aborted, no cycle can exist between the transactions. They do not affect the internal consistency of the database and global serialization is maintained. If either of the transactions is aborted, its changes to the database are nullified and thus no cycle can exist between the transactions. Now let us assume that both transactions will commit.

In order for the global serialization graph $SG(H)$ to have a cycle we must have the following partial serialization graphs: $Tr_1 \rightarrow \cdots \rightarrow Tr_2$ and $Tr_2 \rightarrow \cdots \rightarrow Tr_1$. Since on strict policies transactions cannot reference uncommitted data, we get $c_1 < c_2$ from the first serialization graph, and $c_2 < c_1$ from the second serialization graph. Thus, the global serialization graph has a dependency $c_1 < c_2 < c1$, which leads to $c_1 < c_1$. This is impossible, so one of the transactions Tr_1, Tr_2 must violate local strict policy. Since the assumption is wrong, we can say that the global serialization graph $SG(H)$ is acyclic. $\square$

It is easy to prove that a strict policy on all nodes is a necessary requirement to completely avoid global concurrency control.

Theorem 4.2 *Let us have a distributed database with nodes $N_1 \ldots N_n$. If any node N_i does not follow a strict concurrency control policy, the local transaction policies cannot ensure global serialization.*

Proof: Let us have nodes N_i and N_j $(i \neq j)$ where N_j does not follow a strict policy. Then the following histories are possible:

$$N_i : o_1[x]c_1p_2[x]c_2$$
$$N_j : o_2[y]p_1[y]c_2c_1,$$

where operations o_i and p_i are any conflicting operations.

The serialization graphs we get are $SG_i(H) = Tr_1 \rightarrow Tr_2$, and $SG_j(H) = Tr_2 \rightarrow Tr_1$. Thus the global serialization graph has a cycle: $Tr_1 \rightarrow Tr_2 \rightarrow Tr_1$. $\square$

Since in the previous theorem the only requirement to the operations is that they conflict, the theorem holds regardless of the types of conflicts. Thus, although strict recoverable concurrency control policies are somewhat limiting, they alone can guarantee global transaction serialization from local serialization alone.

A corollary of the previous theorems shows that in order to achieve global isolation between distributed transactions, it is sufficient that distributed database nodes follow any local strict concurrency control policy. This can be seen from the serialization graphs $SG_i(H)$. The graphs are independent of the chosen concurrency control policy as long as the policy is strict. We can get different types of strict transaction histories depending on the chosen strict concurrency control policy. The corollary also allows nodes to use both pessimistic and optimistic strict local concurrency control policies and still maintain global serialization.

Pessimistic concurrency control policies are often strict. This is due to the nature of the policies. When a transaction wants to access a data item, it must first get a lock to it. Most pessimistic policies use hard locks: a lock is allowed only when no conflicting transaction exist. The requirement here is that the lock-holding transaction must keep the locks until it is ready to commit. This ensures that other transactions cannot access data items before the currently holding transaction has committed, and that way transactions may access only committed data. A typical variation of this is called Strict two-phase locking [BHG87].

Yet pessimistic concurrency control policies have another unwanted feature: deadlocks. It is possible that two transactions are waiting for each other to release hard locks even when local concurrency control is strict. Thus, other means are needed to detect deadlocks in a distributed database. Instead of having a global concurrency controller, the database needs a global deadlock detector on each node. Again resources are needed to maintain global transaction management.

Optimistic concurrency control policies are usually not strict. They allow read and write conflicts during transaction execution. It is possible that on node N_1 is a serialization graph $Tr_1 \rightarrow Tr_2$ while on node N_2 the serialization graph is $Tr_2 \rightarrow Tr_1$. Both histories are correct and well possible within optimistic concurrency control. Yet the global graph has a cycle. Due to this using most optimistic concurrency control policies on distributed databases need global concurrency control managers on nodes that maintain global serialization. On the other hand, optimistic concurrency control policies avoid deadlocks since data items are never hard-locked.

Both pessimistic and optimistic concurrency control policies have variations that both allow strict histories and avoid deadlocks. As such they are realistic alternatives to Yalin concurrency control.

The pessimistic concurrency control policy that avoids deadlocks is called Conservative two-phase locking (or Conservative 2PL) [BHG87]. It avoids deadlocks by requiring each transaction to obtain all of its locks before any of its operations are executed. A deadlock cannot exist since it is not possible for a transaction to get more locks after the execution has started. Thus two transactions can never wait for each other to release their locks. Together with Strict 2PL we get a strict pessimistic concurrency control policy that avoids deadlocks.

The optimistic concurrency control policy that avoids global concurrency control is also the most traditional one. The policy goes as follows.

1. When a transaction is started, a local data area is initialized for its use.

2. Whenever a transaction wants to access a data item for reading or writing, the data item is copied to the local data area where the transaction is allowed to modify the data item. A soft lock is set to the data item to keep track of transactions that have accessed this data item.

3. When the transaction is ready to commit, it is first checked if it has a conflict on any of the data items. If the conflict exists, either the committing transaction is aborted or the con- flicting transactions are aborted. If the conflict is with an already committed transaction,

the committing transaction is always aborted.

4. Once the result of the commit is known, the transaction may join into an atomic commit procedure. This finally determines the result of the commit.

5. If the transaction is allowed to commit, the modified data items are written back to the database with one atomic operation.

This policy is strict since transactions are allowed to access only committed data. All modifications are done in local data areas of executing transactions. It is also optimistic since transactions never have to wait for each other to release data items. All conflicts are resolved at commit time. As a result it avoids global concurrency control management and deadlocks.

One of the good sides of optimistic concurrency control is that transactions never block each other. This feature allows real-time databases, such as Yalin, to use different scheduling policies with optimistic concurrency control. In pessimistic concurrency control this requires some kind of priority policy: what to do when a low-priority transaction holds locks that a high-priority transaction wants to get. This problem has usually been solved by either aborting the low-priority transaction or allowing it temporarily to have the same priority as the high-priority transaction. These solutions can be used in Yalin as well.

Nevertheless, an optimistic concurrency control policy is the best alternative for Yalin since it is a real-time database. Transaction deadline management is easier in an optimistic concurrency control policy since the scheduler has full control of transactions. No kind of transaction blocking is possible. Pessimistic concurrency control policies usually behave better in traditional databases.

The traditional optimistic concurrency control policy is a candidate for Yalin concurrency control management. It is both strict and optimistic, and so it fulfills the requirements that we set to the policy. The Strict OCC is a strong algorithm that causes lots of unnecessary restarts. A modified version of the algorithm lowers the number of restarts. When a transaction is marked for abortion due to a conflict with a committing transaction, it is put on a wait state until the voting phase of the committing transaction is over. If the vote result is NO, the waiting transaction is released. Otherwise it is aborted. Also, when a transaction is at the wait state, it can lend uncommitted data to transactions in the optimistic assumption that it will commit. If the transaction is aborted, the borrowing transactions are aborted as well. These optimizations are based on the optimistic atomic commit protocol algorithm PROMPT [GHRS96, HRG00].

4.6 Database global commits

While the local concurrency control, as presented in the previous section, can alone maintain most ACID-properties globally, it cannot guarantee Atomicity. This can easily be seen in the next example.

Let us assume that we have transactions Tr_1 and Tr_2 that both execute on nodes N_1 and N_2. Using strict optimistic concurrency control we can guarantee that any history where the transactions participate is serializable. However, if Tr_1 on node N_1 conflicts with Tr_2 on node N_1 while the transactions do not conflict on node N_2, the transaction Tr_1 (or Tr_2 depending on the commit policy) is aborted on node N_1 while it is not aborted on node N_2. Thus, transaction atomicity is violated even when global serialization is not.

Various atomic commit protocols have been proposed in literature. The most common ones are based on voting. Each participating subtransaction votes whether the transaction should be committed or aborted. Depending on the algorithm some or all subtransactions must vote for commit in order to allow the transaction to commit.

In Yalin we use the traditional two-phase commit protocol as described earlier. It is a very simple and elegant protocol which ensures atomic commitment of distributed transactions. Due to this it is very suitable for Yalin since it is simple to implement yet efficient enough. Its weaknesses are in failure conditions when a transaction is not certain if it should commit or abort. This condition is called a blocking condition since it blocks the transaction from committing or aborting.

The blocking condition in a two-phase commit protocol is due to either a network or site failure. The Yalin network is very reliable since it is built on top of the telecommunications network, and the bus architecture in clusters is almost error-free. Naturally a Yalin node may go down due to a software failure, but even these cases should be rare.

If a two-phase commit protocol is not sufficient, the next alternative is to use the three-phase commit protocol as introduced by Skeen [Ske81]. The protocol adds one extra message phase to the commit processing. The protocol is nonblocking when failures are restricted to site failures [OV99]. As such it suits Yalin well. Unfortunately it also adds new message exchange rounds between the leader and participants so committing is slower than in a two-phase commit protocol.

Distributed transactions in Yalin may execute either at cluster level or at node level, or at both levels. Due to this the Yalin atomic commit protocol is also a two-level commit protocol. The algorithm for this is simple. Let us consider a case where transaction Tr is ready to commit.

1. Set the result to abort or commit depending on the result of the transaction Tr.

2. If Tr is not a subtransaction of a node level transaction, go to 4.

3. Set Tr into a two-phase commit procedure at node level either as a leader or as a participant. Save the result of the node procedure.

4. If Tr is not a subtransaction of a cluster level transaction, go to 6.

5. Set Tr into a two-phase commit procedure at node level either as a leader or as a participant. Save the result of the cluster procedure.

6. Commit or abort the transaction depending on the result.

Unfortunately up to two voting sessions are needed to complete a transaction. This cannot be avoided since cluster and node levels are independent of each other. On the other hand, the two-level architecture allows different protocols to be used at cluster and node levels. Thus, it is possible to use a three-phase commit protocol at cluster level and two-phase commit protocol at node level when necessary.

Chapter 5

Database Analysis

In this chapter we introduce our transaction-based queueing model analysis tools and use them for Yalin performance evaluation. We use the approach presented by Highleyman [Hig89] and derive our specific model equations from it. Then we use our equations to analyze an X.500 Database entry node and a cluster in Yalin.

The transaction types and data volumes have been defined in earlier chapters. This information, along with the Yalin architecture, gives a basis to analyze the system. We wish to find the Yalin bottlenecks. We are especially interested to see how the X.500 Database entry node can accept external requests, how distribution affects transaction execution times, how the cluster architecture behaves, how parallel nodes can co-operate, and what kind of estimates we can make for transaction execution times.

Other interesting areas to consider would be various priority-based scheduling policies and fault tolerance. They are not included in the Yalin queueing model. The scheduling policies are somewhat difficult to analyze with queueing models. The basis of our analysis resides in a non-priority transaction model. Fault tolerance is not analyzed since it is beyond the scope of the thesis. We assume that the bus, the network, and clusters are fault-tolerant.

5.1   General database analysis tools

Analyzing such a complex transaction-based system as Yalin is not a straightforward task. In order to simplify the analysis we define a toolbox of queueing model analysis methods for transaction-based systems.

The analysis toolbox is based on standard Yalin components: transactions, managers, and request arrival rates. We use the components to create a general database analysis tool that is suitable for analyzing Yalin traffic. The same tools can be used to analyze any transaction-based system including a traditional database management system.

Most research in queueing theory has concentrated on very specific queueing models and tools. True enough, general formulas for queueing modeling allow a detailed analysis of a system. The drawback of the detailed approach is that using general formulas requires a good mathematical background, deep analysis, and perhaps a long time to create an appropriate analysis model. In our thesis we take a different approach. We use a simple model where all arriving requests are Poisson-distributed, and the only shared resources are the CPU and servers. These simplifications allow us to generate efficient formulas for solving very complex transaction-based queueing models.

Let us have a set of server nodes N_i each of which has an incoming queue, a set of transaction types Tr_j that arrive to the system, and a set of incoming service request arrival flows f_k that describe how requests traverse between server nodes:

$$N = \{N_1,\ldots,N_n\},$$
$$Tr = \{Tr_1,\ldots,Tr_m\}, \tag{5.1}$$
$$F = \{f_1,\ldots,f_l\}.$$

Each flow f_k has a request arrival rate R_k and an average service time t_k. The request arrival rate prescribes how many service requests arrive to a server in a time interval, and the service time prescribes how long the server needs to serve this kind of request when it has arrived. Thus,

$$f_k = (R_k, t_k), \tag{5.2}$$

where R_k is the request arrival rate for flow f_k and t_k is the average service time for flow f_k requests. Let us mark $f_k.R$ when we refer to element R_k of f_k, and $f_k.t$ when we refer to element t_k of f_k.

The transaction flow logic in the model is presented with *transaction patterns*. A transaction pattern is a route through the analyzed system for a transaction type Tr_j. We have a set of transaction patterns for each transaction type:

$$TP(Tr_j) = \{TP_1(Tr_j), \dots TP_o(Tr_j)\}, \tag{5.3}$$

where one transaction pattern $TP_h(Tr_j)$ is defined as follows:

$$TP_h(Tr_j) = \{(N_i, N_j, f_{ijk}) | N_i, N_j \in N, f_{ijk} \in F\}. \tag{5.4}$$

The incoming flows to a server are divided into a set of transaction patterns, but in the analysis they are calculated together. Thus, we define the incoming flows of transaction type Tr_j for node N_i as follows:

$$F(N_i, Tr_j) = \{f_{ijk} \in F | (A, B, f_{ABk}) \in TP_h(Tr_j), A \in N, B = N_i, TP_h(Tr_j) \in TP(Tr_j)\}. \tag{5.5}$$

The transaction patterns simplify model calculations. Although they are only a way to arrange request arrival rate flows, they give a possibility to analyze the system behavior in smaller parts. In a small or a medium system a few patterns are sufficient. For instance, in Yalin analysis the X.500 Database entry node needs one transaction pattern. Large systems need a lot of possibly complex transaction patterns. For instance, the combined cluster and node analysis of Yalin has six large transaction patterns.

The set of nodes N is divided into a set of clusters. Each cluster models a set of servers that share the same CPU. Some servers are completely hardware driven, for instance DMA-based disks. Such a server forms a cluster alone.

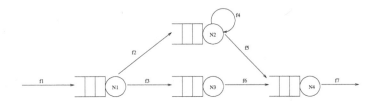

Figure 5.1: Example traffic model.

$$C = \{C_1, ..., C_h\}, C_i = \{N_i | N_i \in N\}. \tag{5.6}$$

The clusters form a disjoint cover to the set N. Thus, each server must belong to exactly one cluster.

$$\cup_i C_i = N, C_i \cap C_j = \emptyset \Leftrightarrow C_i \neq C_j. \tag{5.7}$$

Example 5.1 *Let us consider a simple system that has one transaction type, four servers, and seven flows (Figure 5.1). The first server N_1 receives flow f_1. The flow is divided into two flows f_2 and f_3 which connect to nodes N_2 and N_3. The flow f_4 is a recursive flow in N_2. Flows f_5 and f_6 connect nodes N_2 and N_3 to server N_4, and flow f_7 leaves node N_4.*

The example system has two transaction patterns (Figure 5.2). First, we have a transaction pattern that starts from node N_1, goes to node N_2 where it has zero or more iterations in node N_2. When the iterations are over, the pattern goes to node N_4 where it exits. Second, we have a transaction pattern that starts from node N_1, goes to node N_3 and then to node N_4 where it exits.

The example system had flows f_1 to node N_1 which is present in two transaction patterns. This is not possible since each flow is allowed to be present only in a single pattern. Thus, flow f_1 is divided into two flows f_{1a} and f_{1b}. Similarly flow f_7 is divided into two flows f_{7a} and f_{7b}. The divided flows represent the fact that some percentage of arriving requests take the first transaction pattern while the rest take the second transaction pattern. Thus, the combined flow $f_{1a} + f_{1b}$ to node N_1 is equal to the original flow f_1.

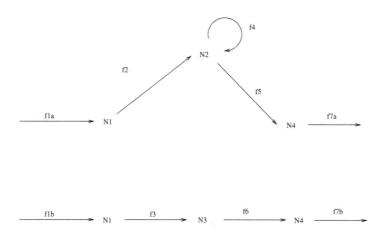

Figure 5.2: Example traffic model transaction patterns.

Formally we have the following example system:

$$N = \{N_1, N_2, N_3, N_4\},$$
$$Tr = \{Tr_1\},$$
$$TP(Tr_1) = \{TP_1(Tr_1), TP_2(Tr_1)\},$$
$$TP_1(Tr_1) = \{(NIL, N_1, f_{1a}), (N_1, N_2, f_2), (N_2, N_2, f_4), (N_2, N_4, f_5), (N_4, NIL, f_{7a})\}$$
$$TP_2(Tr_1) = \{(NIL, N_1, f_{1b}), (N_1, N_3, f_3), (N_3, N_4, f_6), (N_4, NIL, f_{7b}\}.$$

Here NIL states that the server in the connection is missing. It is possible that one of the servers is missing in a transaction pattern element, but not both servers. The NIL connection is syntactic sugar. It only shows that the system is open: transaction requests arrive from outside, and they leave the system as well.

The transaction patterns TP_1 and TP_2 give us the flow information for each node. Following *the definitions in equation 5.5 we get the following flow sets:*

$$F(N_1, Tr_1) = \{f_{1a}, f_{1b}\},$$
$$F(N_2, Tr_1) = \{f_2, f_4\},$$
$$F(N_3, Tr_1) = \{f_3\},$$
$$F(N_4, Tr_1) = \{f_5, f_6\}.$$

Moreover let us assume that servers N_1 and N_4 are hardware driven and do not need CPU services. Servers N_2 and N_3 share the same CPU. This division gives us the following clusters:

$$C_1 = \{N_1\},$$
$$C_2 = \{N_2, N_3\},$$
$$C_3 = \{N_4\}. \quad \square$$

In the transaction patterns of the example we omit queues and server circles. This is only a drawing convention that distinguishes transaction patterns from full queueing model figures.

A transaction pattern may have the same server several times since a transaction pattern describes how control flows in the analyzed system. If we have a pattern that has several paths via the same server then the server appears several times in the transaction pattern. The cluster transaction patterns in Section 5.4 are examples of this.

For each incoming service request, the total time T in a server is the sum of the queue waiting time Q and the service processing time S. The service processing time S is divided into two components: CPU dispatch time D and service execution time E. Thus, we get

$$T = Q + S,$$
$$S = E + D. \qquad (5.8)$$

The queue waiting time Q depends on the average utilization L of the server and the remaining service time kS of the currently active transaction in the server. The formula, as in [Hig89], is

$$Q = \frac{L}{1-L} kS, \qquad (5.9)$$

where k depends on the distribution of an arriving request. Usually the distribution is Poisson distribution, in which case $k = 1$. That is, on average the remaining time of the currently processed request is equal to the average time in the server. From now on we assume that arriving requests are Poisson distributed: we set $k = 1$. Thus, we get

$$Q = \frac{L}{1-L} S.$$

(5.10)

The utilization L tells how much the server is occupied. It is defined as

$$L = RS,$$

(5.11)

where R is the incoming request rate and S is the service processing time in the server.

Now let us analyze the traffic model in a node N_i for transaction types in Tr. The node N_i has incoming Tr_j type requests from l sources. When all these are summarized, we get the request arrival rate for transaction type Tr_j in node N_i. When we summarize over all transaction types, we get the total arrival rate to node N_i:

$$R_i = \sum_{j=1}^{m} \sum_{k=1}^{l} f_k(N_i, Tr_j).R,$$

(5.12)

where $f_k(N_i, Tr_j).R$ is the arriving rate to node N_i from transaction type Tr_j and source flow f_k.

The total service time T_i in node N_i is

$$T_i = Q_i + S_i,$$

(5.13)

as earlier.

The queue waiting time Q_i depends on the node utilization L_i and the service processing time S_i. The formula is the same as earlier:

$$Q_i = \frac{L_i}{1-L_i} S_i.$$

(5.14)

The service processing time S_i on node N_i is the sum of the service execution time E_i and the CPU dispatch time D_i. This gives us:

$$S_i = E_i + D_i.$$

(5.15)

The service execution time E_i on node N_i is a weighted average of all execution times from different request sources. The weighting factor here is the request arrival rate. Thus, we get:

$$E_i = \frac{\sum_{j=1}^{m}\sum_{k=1}^{l}(f_k(N_i,Tr_j).R)(f_k(N_i,Tr_j).t)}{R_i},\qquad(5.16)$$

where $f_k(N_i,Tr_j).t$ is the average execution time in node N_i from transaction Tr_j and source flow f_k.

Let us define

$$w_i = \sum_{j=1}^{m}\sum_{k=1}^{l}(f_k(N_i,Tr_j).R)(f_k(N_i,Tr_j).t).\qquad(5.17)$$

Then we can write

$$E_i = \frac{w_i}{R_i}.\qquad(5.18)$$

When a service process receives a service request, it must first wait in the CPU queue to gain access to the CPU. This wait time is modeled with the parameter D. We will assume that the CPU is in cluster $C_l \in C$. Here we assume that the cluster has at least two servers. The more servers use the same CPU, the better the Poisson distribution estimation models the actual arrival rates.

We already know that the wait time in a queue is

$$Q = \frac{L}{1-L}S,$$

$$L = RS.$$

In the CPU dispatch case, R is the sum of the arrival rates in cluster C_c. The average execution time d is the weighted average of the service execution times in the cluster, excluding dispatch time. With this information we get:

$$D_i = d_c = \frac{l_c}{1-l_c}s_c,\qquad(5.19)$$

where l_c is the CPU utilization on cluster C_c, and s_c is the average CPU service time on cluster C_c.

Let us define

$$u_c = \sum_{i=1}^{c_i} \sum_{j=1}^{m} \sum_{k=1}^{l} (f_k(N_i, Tr_j).R)(f_k(N_i, Tr_j).t) \qquad (5.20)$$

and

$$R_c = \sum_{i=1}^{c_i} R_i. \qquad (5.21)$$

Then we get

$$l_c = s_c \sum_{i=i}^{c_i} R_i = s_c R_c, \qquad (5.22)$$

and

$$s_c = \frac{\sum_{i=1}^{c_i} \sum_{j=1}^{m} \sum_{k=1}^{l} (f_k(N_i, Tr_j).R)(f_k(N_i, Tr_j).t)}{\sum_{i=i}^{c_i} R_i} = \frac{u_c}{R_c}. \qquad (5.23)$$

Moreover, u_c may be written into a simpler form.

$$u_c = \sum_{i=1}^{c_i} w_i. \qquad (5.24)$$

As a result, we get the following formula for the CPU dispatch time d_c on cluster C_c:

$$D_i = d_c = \frac{R_c(\frac{u_c}{R_c})^2}{1 - R_c(\frac{u_c}{R_c})} = \frac{u_c^2}{R_c(1 - u_c)}. \qquad (5.25)$$

The CPU utilization on cluster C_c is as follows:

$$l_c = s_c R_c = \frac{u_c}{R_c} R_c = u_c. \qquad (5.26)$$

A special case occurs to clusters with only one server node. Since the requests arrive only from one source, the distribution is no longer Poisson distributed. In fact, as long as the server has a queue we do not need a CPU dispatch time. Thus, in this special case $d_c = 0$.

When we combine the previous formulas, we get the following utilization and service times for node N_i:

$$T_i = Q_i + E_i + D_i = \frac{L_i}{1 - L_i}(E_i + D_i) + (E_i + D_i) = (\frac{L_i}{1 - L_i} + 1)(E_i + D_i). \qquad (5.27)$$

Furthermore we get

$$L_i = R_i S_i = R_i \left(\frac{w_i}{R_i} + \frac{u_c^2}{R_c(1 - u_c)} \right) = w_i + \frac{R_i}{R_c} \frac{u_c^2}{1 - u_c}. \tag{5.28}$$

Finally this gives us the service time T_i:

$$T_i = \left(\frac{L_i}{1 - L_i} + 1 \right) \left(\frac{w_i}{R_i} + \frac{u_c^2}{R_c(1 - u_c)} \right). \tag{5.29}$$

The service time on node N_i for transaction Tr_j is as follows:

$$T_{ij} = Q_i + E_{ij} + D_i = \frac{L_i}{1 - L_i}(E_i + D_i) + E_{ij} + D_i. \tag{5.30}$$

The only new variable in the previous formula is E_{ij}. It is the average execution time for transaction Tr_j on node N_i. It is a weighted average of all incoming execution times for transaction Tr_j:

$$E_{ij} = \frac{\sum_{k=1}^{q_{ij}} (f_k(N_i, Tr_j).R)(f_k(N_i, Tr_j).t)}{\sum_{k=1}^{l} f_k(N_i, Tr_j).R}. \tag{5.31}$$

Let us define

$$w_{ij} = \sum_{k=1}^{l} (f_k(N_i, Tr_j).R)(f_k(N_i, Tr_j).t), \tag{5.32}$$

and

$$R_{ij} = \sum_{k=1}^{l} f_k(N_i, Tr_j).R. \tag{5.33}$$

With these definitions we get the final form of the average service time for transaction Tr_j on node N_i.

$$T_{ij} = \frac{L_i}{1 - L_i} \frac{w_i}{R_i} + \frac{w_{ij}}{R_{ij}} + \left(\frac{L_i}{1 - L_i} + 1 \right) \frac{u_c^2}{R_c(1 - u_c)}. \tag{5.34}$$

5.1.1   Summary of notations and formulas

The 34 equations above define both the middle and the final results of our toolbox. Here we summarize the final results: toolbox notations and formulas. First, we describe our notations:

N_i : A server (node) in the system.

N : A set of nodes in the system.

Tr_j : A transaction type in the system.

Tr : A set of transaction types in the system.

f_k : A service request flow in the system.

F : A set of service request flows in the system.

$f_k(N_i, Tr_j)$: Flow f_k to node N_i of transaction type Tr_j.

$f_k(N_i, Tr_j).R$: Request arrival rate for flow f_k above.

$f_k(N_i, Tr_j).t$: Average service time for flow f_k above.

$TP(Tr_j)$: A set of transaction patterns for transaction type Tr_j.

$TP_h(Tr_j)$: One transaction pattern for transaction type Tr_j.

$F(N_i, Tr_j)$: Node N_i service request flows of transaction type Tr_j.

C : A set of clusters.

R_i : Node N_i request arrival rate.

w_i : Node N_i average service time multiplier.

E_i : Node N_i service execution time.

R_c : Cluster C_c request arrival rate.

u_c : Cluster C_c service time multiplier.

l_c : Cluster C_c utilization.

d_c : Cluster C_c CPU execution time.

D_i : Cluster C_c dispatch time.

s_c : Cluster C_c average service time.

L_i : Node N_i utilization.

T_i : Node N_i total time.

w_{ij} : Node N_i service time multiplier for transaction type Tr_j.

R_{ij} : Node N_i request arrival rate for transaction type Tr_j.

T_{ij} : Node N_i total time for transaction type Tr_j.

Using the notations above we summarize our formulas in equations 5.35 and 5.36:

$$N = \{N_1, \ldots, N_n\} \cup \{NIL\}$$

$$Tr = \{Tr_1, \ldots, Tr_m\}$$

$$F = \{f_1, \ldots, f_l\},$$

$$f_k = (R_k, t_k),$$

$$TP(Tr_j) = \{TP_1(Tr_j), \ldots TP_o(Tr_j)\},$$

$$TP_h(Tr_j) = \{(N_i, N_j, f_{ijk}) | N_i, N_j \in N, f_{ijk} \in F\},$$

$$F(N_i, Tr_j) = \{f_{ijk} \in F | (A, B, f_{ABk}) \in TP_h(Tr_j), A \in N, B = N_i, TP_h \in TP(Tr_j)\},$$

$$C = \{C_1, \ldots, C_h\}, C_i = \{N_i | N_i \in N\},$$

$$R_i = \sum_{j=1}^{m} \sum_{k=1}^{l} f_k(N_i, Tr_j).R,$$

$$w_i = \sum_{j=1}^{m} \sum_{k=1}^{l} (f_k(N_i, Tr_j).R)(f_k(N_i, Tr_j).t), \tag{5.35}$$

$$E_i = \frac{w_i}{R_i},$$

$$R_c = \sum_{i=1}^{c_i} R_i,$$

$$u_c = \sum_{i=1}^{c_i} w_i,$$

$$l_c = u_c,$$

$$d_c = \frac{u_c^2}{R_c(1 - u_c)},$$

$$D_i = d_c,$$

$$s_c = \frac{u_c}{R_c}.$$

$$L_i = w_i + \frac{R_i}{R_c} \frac{u_c^2}{1 - u_c},$$

$$T_i = (\frac{L_i}{1 - L_i} + 1)(\frac{w_i}{R_i} + \frac{u_c^2}{R_c(1 - u_c)}),$$

$$w_{ij} = \sum_{k=1}^{l} (f_k(N_i, Tr_j).R)(f_k(N_i, Tr_j).t), \qquad (5.36)$$

$$R_{ij} = \sum_{k=1}^{l} f_k(N_i, Tr_j).R,$$

$$T_{ij} = \frac{L_i}{1 - L_i} \frac{w_i}{R_i} + \frac{w_{ij}}{R_{ij}} + (\frac{L_i}{1 - L_i} + 1)\frac{u_c^2}{R_c(1 - u_c)}.$$

5.2 Example: Traditional DBMS analysis

In order to better understand the analysis tools, we will analyze a traditional database management system from both an optimistic and a pessimistic concurrency control point of view. The
queueing model of the system can be seen in Figure 5.3.

In the example we want to compare optimistic and pessimistic concurrency control along
with information of various database processes and possible bottlenecks. This information is
best analyzed with a single transaction type and a transaction pattern. Naturally in an actual
case several types of transactions may exist. We use an average transaction for calculations.
The results are as accurate as they would be with several types of transactions.

For the transaction type Tr, we have an incoming request arrival rate R to the system. Each
request arrives to the incoming network server (Net In). From there it is forwarded to the
transaction manager (TM) that creates a new transaction.

The new transaction is sent to the scheduler which schedules transaction operations. The
total number of operations in the transaction is $n + m$, where n is the number of operations that
need to access database data, and m is the number of operations that do some calculation with
received data and thus need only the CPU.

The cache manager (CM) manages database data buffers. When an operation can be served
directly in the buffers, the operation is executed immediately and the result is sent back to the

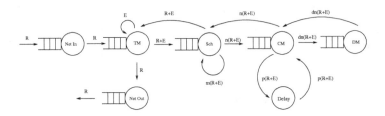

Figure 5.3: DBMS traffic model.

transaction. If the operation cannot be served with memory buffers alone, the cache manager consults the disk manager to physically access database disks. This happens with a probability d.

For each transaction, the probability for a conflict is p. The conflict management depends on whether we have an optimistic or a pessimistic concurrency control method.

In the optimistic concurrency control, the conflicts are resolved when a transaction is committing. In this model we assume that the committing transaction is aborted and restarted if it has conflicts.

The aborted transaction rate E depends on the probability that the transaction Tr has conflicts with other transactions. This occurs with a probability p which depends on the read and write sets of the transactions.

For each transaction Tr, the probability of a conflict is p. Thus, in the first round we have Rp conflicts, in the second round Rp^2 conflicts and so on. The total number of conflicts and restarts is

$$E = R \sum_{i=1}^{\infty} p^i = R \frac{p}{1-p}. \tag{5.37}$$

In the pessimistic concurrency control, on each CM operation the transaction may be blocked when it wants to lock a data item that has been locked by another transaction. Since the conflict probability for the transaction is p, the probability on a certain round is p/n. If the conflict occurs, the transaction is put into a wait state until the conflicting transaction releases the blocking lock. This occurs after the blocking transaction has finished.

Since the arrival rates to the system are Poisson distributed, the average waiting time for the blocking transaction is equal to the average execution time for a transaction Tr. After that the transaction is released. In the queueing model this is modeled with a node Delay. A transaction waits in the node for the average transaction execution time. The node Delay does not have a queue.

The average transaction execution time is a sum of the execution times on each node. The execution times depend on how many times a transaction will visit each node and how long it will spend on them. Thus, we get:

$$T = T_{\text{N-I}} + T_{\text{TM}} + \ldots + T_{\text{N-O}} + \frac{pn}{n}T \Leftrightarrow T - pT = T_{\text{N-I}} + \ldots \quad (5.38)$$

As a result we get the following total time equation:

$$T = \frac{T_{\text{SUM}}}{1-p}, \quad (5.39)$$

where T_{SUM} is a sum of the service times excluding the delay time.

In addition to the previous managers, a pessimistic concurrency control database management system has a manager that detects deadlocks. In the model we have modeled the Cache manager to include its functionality. Whenever a CM operation occurs, it is also tested against deadlocks. This can be done with a deadlock detection graph or a similar algorithm. If a deadlock occurs, one of the deadlocking transactions is aborted.

The probability that a deadlock is detected depends on the probability of a conflict. We take a simple approach where the probability of a deadlock is p^2, where p is the probability of a conflict. This comes from an extremely pessimistic assumption that when a transaction conflicts, any of the possibly conflicting transactions may have an ongoing conflict with probability p. Since the transactions are independent, the total conflict probability is p^2.

Whenever a deadlock is detected, one of the transactions must be aborted to solve the deadlock. Thus also in a pessimistic concurrency control the restart rate E exists. We can use the same formula as in the optimistic concurrency control, except that the probability of a restart is

p^2. This gives us

$$E = R \sum_{i=1}^{\infty} (p^2)^i = R \frac{p^2}{1 - p^2}. \tag{5.40}$$

The average execution times t_{ijk} for different incoming flows depend on the current hardware and database management system software. We do not get into such details here. Instead, on each node N_i we consider a fixed execution time t_i for all incoming flows on node N_i. The times in the example are based on modern hardware and software configurations but nevertheless they are only estimates.

The Net In and Net Out have very simple algorithms. The Net In accepts requests and forwards them to the TM. The Net Out accepts replies from TM and forwards them to the clients. We consider them to take an average 0.01ms for service processing.

The TM creates new transactions and commits or aborts finished ones. These tasks are relatively simple so we consider them to need 0.1ms.

The Scheduler mostly decides which transaction is executed next. This is a straightforward task which needs 0.1ms. However, with probability $m/(n+m)$ the transaction needs only CPU access, at which time no other transactions can be scheduled (since they all use the same CPU). We consider the execution to take an average 1ms time. This gives us the average execution time $(n * 0.1ms + m * 1.0ms)/(m+n)$.

The Cache manager must check if a requested data item is in the cache, manage locking and possibly forward the request to the disk manager. This is not a straightforward task since data management must be taken into account. We consider this to take 1ms.

The Disk manager is a special manager since it offers a hardware connection to the disks. A disk access time can be estimated with the following formula:

$$t_{\text{disk}} = t_{\text{seek}} + t_{\text{delay}} + t_{\text{access}}, \tag{5.41}$$

where

$$t_{\text{delay}} = \frac{1}{2v_{\text{rotation}}/60}, \tag{5.42}$$

and

$$t_{\text{access}} = v_{\text{access}} * s_{\text{block}}. \tag{5.43}$$

Here t_{seek} is the average seek time, $v_{rotation}$ is the disk rotation speed (rounds/minute), v_{access} is the average access time (bytes/second), and s_{block} is the block size in bytes.

The disk read/write time depends on the data transfer rate, disk seek time, disk rotation speed, and the size of the disk sector. A very good and efficient hard disk can handle a 20MB/s read and write access, a 4ms seek time, and a 7200 rounds/minute rotation time. We consider a disk sector to be 4096 bytes long. These values give us

$$t_{disk} \approx 8.4\text{ms}. \tag{5.44}$$

The Deadlock detector in the pessimistic concurrency control checks the deadlock graph for possible cycles. At full scale this is an extremely difficult task since the cycles can be of arbitrary length. However, since we gave a pessimistic value to the deadlock probability, we can compensate it by giving an optimistic value to the deadlock detector algorithm. We use 1ms for deadlock detection.

Finally, we have to consider the CPU clusters of the model. The Net In, Net Out, and DM are clearly hardware-driven with direct memory access to memory structures. Thus, each of them forms a cluster. Since each cluster has only one server, there is no need to consider dispatch rates.

The rest of the servers, that is TM, Sch, and CM share the same CPU. They form a cluster in the analysis. The pessimistic concurrency control lock delay is not an actual manager so it does not have a cluster. It affects the TM, Sch, and CM CPU utilization since waiting transactions are eventually activated to the CM.

With this information we can compare optimistic and pessimistic concurrency control. The variables to consider are the conflict probability p and disk cache miss probability d. The varying parameter is the request arrival rate R. The average number of CM operations in a transaction is $n = 10$. The average number of calculation operations is $m = 10$.

We will start our analysis from a case, where both the disk cache miss probability and the conflict probability are zero. This is a clearly optimistic case so optimistic concurrency control should behave well. The average execution times can be seen in Figure 5.4.

As we can see from the figure, pessimistic concurrency control can manage about 26 re-

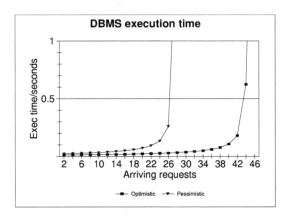

Figure 5.4: Average transaction execution time: $p = 0$, and $d = 0$.

quests/second while optimistic can handle over 40 requests/second. The result is natural since pessimistic concurrency control must still check for deadlocks even when none exist. The difference can clearly be seen in Figure 5.5 where optimistic and pessimistic concurrency control manager utilizations are described.

In optimistic concurrency control, the Scheduler is the first manager to overload. The Cache manager and Transaction manager have clearly lighter utilizations. This is reasonable since the Scheduler also processes transaction operations that do not need database data. The disk manager naturally has no utilization since the disk cache miss probability is zero.

Also in pessimistic concurrency control, the Scheduler is the first manager that overloads. However, here the Cache manager overloads much faster than in the optimistic case. This is due to the deadlock detection that occurs in the cache manager. The extra CPU utilization affects both Scheduler and Cache manager as well. In fact the Scheduler in pessimistic concurrency control receives less requests than in optimistic concurrency control, since the transaction restart rate is smaller in pessimistic concurrency control than in optimistic concurrency control. Yet the pessimistic manager overloads faster than the optimistic one.

The figures are quite similar when we change the disk cache miss and conflict probabilities to 0.1. The average service time can now be seen in Figure 5.6.

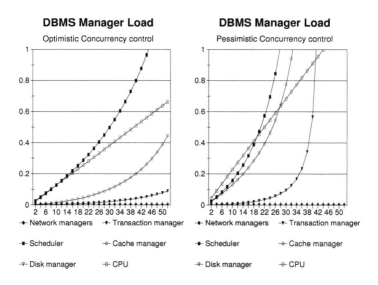

Figure 5.5: Concurrency control manager utilizations: $p = 0$, and $d = 0$.

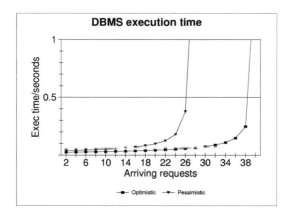

Figure 5.6: Average transaction execution time: $p = 0.1$, and $d = 0.1$.

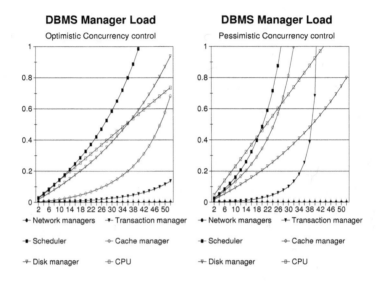

Figure 5.7: Concurrency control manager utilizations: $p = 0.1$, and $d = 0.1$.

The optimistic concurrency control is somewhat affected by the higher probabilities. The effect is on transaction restarts that stress the Scheduler more than in the previous case. Yet the optimistic concurrency control still behaves better than the pessimistic one. However, the pessimistic concurrency control can handle as many requests as in the previous case. This is due to the fact that the bottleneck is still the Scheduler which is affected only by the few restarted transactions. The disk manager overloads much slower and thus does not affect the system performance. In both optimistic and pessimistic cases, the disk manager overloads slower than the Scheduler. Thus, it does not affect the request throughput pace that much (Figure 5.7).

In the previous cases the disk utilization is so small that dominating managers are all software based. However, when the conflict probabilities rise to 0.25, the situation changes. This gives traditional curves in Figure 5.8. Here the conflict probability is $p = 0.25$ and the disk cache miss probability is $d = 0.25$. At small arrival rates the optimistic concurrency control gives slightly better throughput since the software managers still dominate. At around 15 arrivals/second the disk manager starts to dominate. The optimistic concurrency control receives

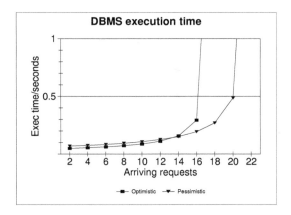

Figure 5.8: Average transaction execution time: $p = 0.25$, and $d = 0.25$.

more disk requests due to a higher restart rate. Thus, it also overloads faster.

The same result can be seen in Figure 5.9. Both in the optimistic and in the pessimistic concurrency control the first manager to overload is the disk manager, but in the pessimistic case the overload happens a little slower.

The same trend would continue when the disk cache miss probability becomes higher. At the same time the software advantages of optimistic concurrency control become negligible compared to the disk cache miss advantages. Thus, curves like those in Figure 5.8 are expected in such cases.

On the other hand, when the disk cache miss probability is small, the optimistic concurrency control can tolerate quite high restart rates and still dominate over pessimistic concurrency control. For instance in Figure 5.10 the conflict probability is $p = 0.5$ and the disk cache miss probability $d = 0.1$. Here the optimistic concurrency control still behaves better than the pessimistic one. The restart rate of optimistic concurrency control is higher than in pessimistic concurrency control, but the extra overhead of deadlock management and lock waits compensates this effectively.

As a result of this example analysis, it appears that pessimistic concurrency control is a suitable concurrency control policy in database management systems that have a medium or large

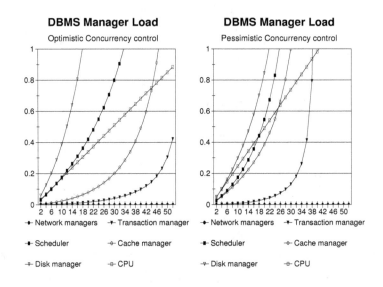

Figure 5.9: Concurrency control manager utilizations: $p = 0.25$, and $d = 0.25$.

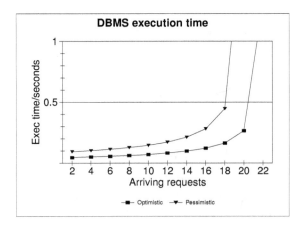

Figure 5.10: Average transaction execution time: $p = 0.5$, and $d = 0.1$.

numbers of disk operations. The disk manager becomes the bottleneck in such systems, and a pessimistic concurrency control needs less disk access due to a smaller restart rate. However, the analysis shows that in database management systems with large memory buffers, and especially in main memory databases, an optimistic concurrency control policy may in fact be better than a pessimistic concurrency control policy.

5.3   X.500 Database entry node analysis

As we described in Chapter 3, the X.500 Database entry node mainly accepts external requests from other networks. It offers an X.500 Directory System Agent (DSA) interface for network clients to the database. The entry may receive three types of operations: X.500 DSA requests, X.500 updates from other DSAs that are in Cluster entry nodes, and X.500 responses that arrive from clusters. The X.500 updates and responses are secure and can proceed immediately. The X.500 requests are first tested in the security manager.

The behavior of the X.500 Database entry node depends on the type of the received operation. If the operation is an X.500 request, it is interpreted and the result is forwarded to one of the Cluster entry nodes in Yalin. If the operation is an X.500 response, it is forwarded immediately to the requester. If the operation is an X.500 update, the directory database is updated. The actual database operations occur in the nodes. Only the results are forwarded back to the X.500 which further forwards them to the request sources.

Since we have three types of requests to the X.500 Database entry node, we also have three types of transactions:

- Tr_1 : X.500 DSA requests,

- Tr_2 : X.500 updates, and

- Tr_3 : X.500 responses.

The architecture of the entry node can be seen in Figure 5.11. It consists of the following managers: the Incoming network request manager (Net In) that accepts new X.500 based transaction requests, the Security manager (Security) that verifies requests from untrusted sources,

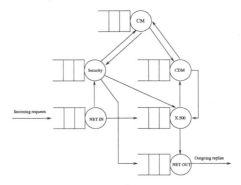

Figure 5.11: X.500 Database entry node model.

the X.500 interpreter (X.500) that interprets requests to internal database management language, the Cluster directory manager (CDM) that maintains directory information about X.500 tree structure, the Disk manager (DM) that offers disk read and write services to Security and CDM servers, and the Outgoing network reply manager (Net Out) that sends replies back to the requesting sources.

The clusters C_i in the X.500 are {Net In}, {Security, X.500, CDM}, {DM}, and {Net Out}. The Security, X.500, and CDM are software managers which share the same CPU. The rest are hardware managers with DMA access to main memory.

We analyze the X.500 Database entry node architecture with a single transaction pattern. When a request arrives, it is accepted into the Net In manager. From there it is sent to the Security manager if the request is not from a reliable source. This occurs with probability p_{j1} which depends on the type of transaction Tr_j. If the request is from a reliable source, it is sent directly to the X.500 interpreter for interpretation.

When a transaction arrives to the Security manager, it has an identity code and a password. The Security manager verifies these and fails access with probability p_{j2}. The Security manager may need to read identification data from the disk. This happens with probability p_{j3}, in which case the Security manager consults the DM.

Once the identity and password of the incoming request have been verified, the request is

seek	Disk seek time in seconds	0.008000s
delay	Disk rotation delay	0.004167s
access	Disk access time bytes/second	25MB/s
block	Disk block size	4096 bytes

Table 5.1: Disk parameters

either accepted or rejected. If it is accepted, it is sent to the X.500 interpreter. If it is rejected, the rejection status is sent to the Net Out which forwards it back to the requesting source.

The X.500 interpreter needs n_j rounds to the CDM to interpret the request. On each round the CDM may need disk access with probability p_{j4}. In such a case it consults the DM.

Once the interpretation in the X.500 manager is over, the interpreted result is sent to the Net Out. The Net Out either forwards the interpreted result to an appropriate cluster, if it was an X.500 request, or returns it back to the sender if it was of any other type.

The service times in the X.500 follow the same pattern as in the traditional DBMS analysis. Net In and Net Out are simple managers that mostly forward requests to other managers. We consider them to take 0.01ms. The Security manager needs only to confirm the incoming request identity and password. We consider this to take 0.1ms time. The X.500 interpreter needs both to call the CDM for directory data and to do the interpretation. We consider this to take 1ms on each round. This time includes interpretation, and result validation. The CDM is a directory manager. As long as the directory data is in the main memory, accessing and validating it is a straightforward task. We consider this to take 0.1ms.

Finally, the DM is a hardware manager for disks. This was analyzed in the traditional database management system example. The values used here are based on current state-of-the-art hard disk values for personal computers. We use a 4096 bytes/block block size, 25MB/second disk access, 7200 rounds/minute disk rotation, and 8ms disk seek time. The values and calculated intermediate results are summarized in Table 5.1.

With these values we get

$$t_{\text{disk}} \approx 12.3\text{ms}. \tag{5.45}$$

The received disk access time for DM is larger than the DBMS example value since here we use a more traditional disk seek time 8ms. These values are equivalent to good conventional hard disk access times while earlier we used values for specialized database management disks.

In the analysis we use the following parameters:

- Incoming transaction probability: $Tr_1 : 0.4, Tr_2 : 0.2, Tr_3 : 0.4$

- Probability that a service request is sent to the Security manager for authentication (p_{j1}): $Tr_1 : 1, Tr_2 : 0, Tr_3 : 0$. Only the transaction type Tr_1 needs authentication.

- Probability that an illegal service request arrives and thus the authentication fails (p_{j2}): $Tr_1 : 0.05. Tr_2 : 0, Tr_3 : 0$.

- Number of X.500 $\rightarrow$ CDM $\rightarrow$ X.500 rounds: $Tr_1 : 10, Tr_2 : 1, Tr_3 : 0$. The interpretation from X.500 to Yalin internal data model is not a straightforward task so we assume ten rounds here. The X.500 directory update needs only one CDM access. The X.500 reply does not need CDM access; the result is returned back to the request source.

- Probability that Security manager needs disk access (p_{j3}): $Tr_1 : 0.1, Tr_2 : 0, Tr_3 : 0$. Only the transaction type Tr_1 needs security manager services.

- Probability that CDM needs disk access (p_{j4}): $Tr_1 : \text{varies}, Tr_2 : 1, Tr_3 : 0$.

The X.500 Database entry node is an easy element to analyze. Even with small disk access probabilities, the disk manager becomes a bottleneck. This can be seen in Figure 5.12 where the disk access probability for transaction type Tr_1 is 0.1. With that disk access probability the system can accept about 60 requests in a second.

The first software manager to overload is the X.500 interpreter. Yet it can tolerate much higher transaction arrival rates than the disk manager. The Cluster directory manager utilization grows slowly at first but on higher arrival rates it can also become a bottleneck. However, with these values the transaction arrival rates can grow very high. When disk manager utilization remains low, the other managers behave well enough to accept almost a hundred requests in a second. This is sufficient to a very large IN/GSM network.

X.500 manager utilization

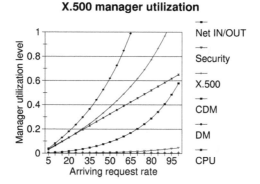

Figure 5.12: X.500 manager utilizations: Tr_1 disk access probability $p_{14} = 0.1$.

The same trend continues when the disk access probability raises. An example of this can be seen in Figure 5.13 where the disk access probability is 0.5. The disk manager is clearly a bottleneck here, but still the system can accept about 15 requests in a second. If this is not enough, the disk access probability must be lowered or extra disks must be included. Adding a disk for the security manager does not help much since only 40 % of arriving requests go via the security manager, and at most one disk access round is needed in a request.

The scenario does not change much if we vary the number of rounds that is needed to interpret a Tr_1 request to Yalin transaction language. The disk manager remains the bottleneck.

5.4   Cluster analysis

The cluster architecture, as seen in Figure 5.14, is a parallel database architecture. It consists of a cluster entry node and a set of database nodes. Transaction requests arrive to the cluster entry node and replies are sent back to the request source via it. Depending on the nature of the request, six transaction types are possible: a local transaction, a bus-level master transaction, a subtransaction within a bus, a cluster-level master transaction, a cluster-level subtransaction, and an X.500 reply/update. Each of the transaction types has a transaction pattern.

The queueing model of the cluster entry node and one of the parallel database nodes can

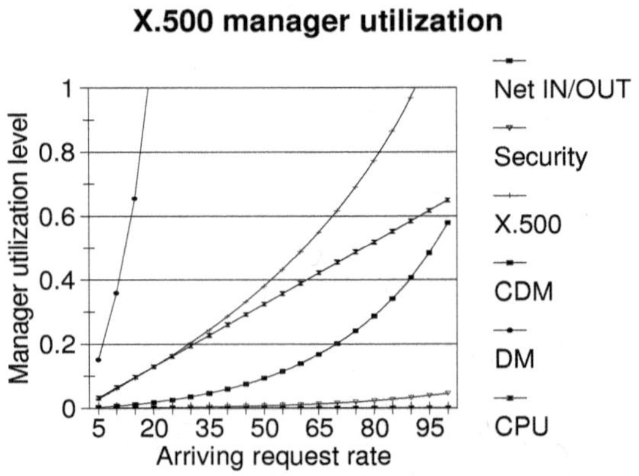

X.500 manager utilization

Net IN/OUT

Security

X.500

CDM

DM

CPU

Manager utilization level

Arriving request rate

Figure 5.13: X.500 manager utilizations: Tr_1 disk access probability $p_{14} = 0.5$.

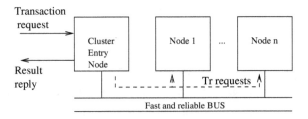

Figure 5.14: Cluster architecture.

be seen in Figure 5.15. In the architecture we have n_c clusters, each of which has n_0 parallel database nodes. Both the network between clusters and the bus between parallel database nodes is modeled as well.

5.4.1 Cluster transaction patterns

The Yalin cluster queueing model, as presented in Figure 5.15, is complex. It consists of 17 managers, a bus, and a network. It has several execution paths through the system. In order to simplify the process of analyzing the queueing model, we use six transaction patterns that describe the transactions of the system. The cluster transaction patterns are as follows: local transaction, master bus-level transaction, slave bus-level transaction, master cluster-level transaction, slave cluster-level transaction, and X.500 service reply or directory update. The patterns are analyzed in turn.

We use constants n_i to describe various request arrival rate multipliers in transaction patterns, and probabilities p_j to describe probabilities for certain event occurrences. Each transaction pattern has its own constants and probabilities that are named locally within a pattern. For instance, the average number of non-database operations (named n_1) in the local transaction pattern may be different from the average number of non-database operations (also named n_1) in the master bus-level transaction pattern case. The names and descriptions are not ambiguous since transaction patterns define a local name space.

Local transaction pattern

The first transaction pattern describes a transaction that is local to a single node. It can be seen in Figure 5.16. This is the most common transaction pattern.

The Net In accepts a request and sends it to the Local call manager (LCM). The LCM translates the request and forwards it to the appropriate node. This happens via the Bus Out manager that is an entry point to the local bus. The requests are divided to all nodes in the bus. In the analysis we assume that the requests distribute evenly to all parallel database nodes in the cluster. The number of arriving requests to a node N_i is R/n_0, where n_0 is the number of nodes

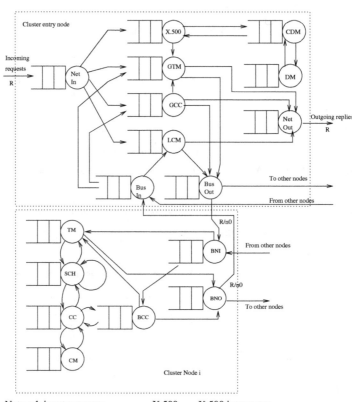

Net In	Network input manager	X.500	X.500 interpreter
CDM	Cluster data manager	DM	Cluster disk manager
GTM	Global transaction manager	Net Out	Network output manager
GCC	Global concurrency controller	LCM	Local call manager
Bus In	Bus input manager (entry)	Bus Out	Bus output manager (entry)
BNI	Bus node input manager (nodes)	BNO	Bus node output manager (nodes)
TM	Transaction manager	SCH	Scheduler
CC	Concurrency controller	CM	Cache manager
BCC	Bus concurrency controller		

Figure 5.15: Cluster queueing model.

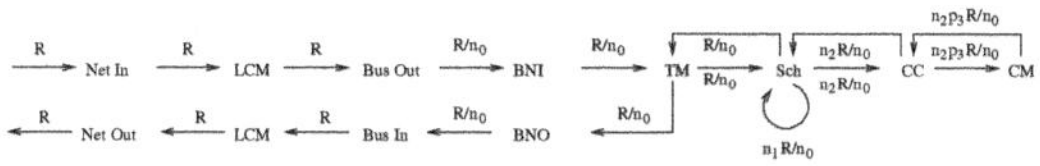

R: Incoming request arrival rate

n_0:   Number of nodes in a cluster

n_1:   Number of CPU-operation rounds

n_2:   Number of CC-operation rounds

p_3:   Probability for a CC-operation to need CM-services

Figure 5.16: Local transaction pattern.

in the cluster and R is the local transaction request arrival rate in the system.

In the local node the Bus node input manager (BNI) accepts the request and sends it to the Transaction Manager (TM). The TM creates a transaction which it sends to the Scheduler. The Scheduler executes the transaction operations. The total number of transaction operations is $n_1 + n_2$ where n_1 is the number of calculation operations executed in the Scheduler, and n_2 is the number of operations forwarded to the Concurrency controller (CC).

In the CC the database operation may need disk access. This happens with probability p_3. If disk access is needed, a request is sent to the Cache Manager (CM) which executes it. The result is then forwarded back to the CC which returns it to the Scheduler. The CM thus receives $n_2 p_3 R / n_0$ requests.

Once the transaction is complete, its status is tested in the TM. If the transaction has no illegal conflicts, it is allowed to commit. Otherwise it is aborted. In both cases the TM sends the result via the Bus node output manager (BNO) to the bus, where cluster entry node Bus In receives it. The Bus In forwards the result back to the LCM which returns it to the Net Out manager. The Net Out manager sends the result back to the transaction requester.

Master bus-level transaction pattern

The second transaction pattern describes a transaction that is bus-distributed to several parallel database nodes in a cluster. In the pattern our node is a master in a transaction. The pattern can

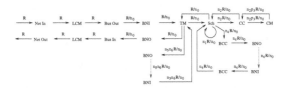

R: Incoming request arrival rate

n_0:   Number of nodes in a cluster

n_1:   Number of CPU-operation rounds

n_2:   Number of CC-operation rounds

p_3:   Probability for a CC-operation to need CM-services

n_4:   Number of bus-level BCC operation rounds

n_5:   Number of master bus-level commit protocol message rounds

n_6:   Average number of slave nodes in a bus transaction

Figure 5.17: Master bus-level transaction pattern.

be seen in Figure 5.17.

The first visited managers are similar to the local transaction pattern. The Net In manager receives a request which is eventually forwarded to the database node N_i. In the node N_i the TM creates a transaction and sends it to the Scheduler for scheduling. Now the transaction has three types of operations: n_1 non-database operations, n_2 local database operations, and n_4 bus-level database operations.

The non-database operations and the local database operations are similar to the previous pattern. The CM access probability is p_3. When a bus operation occurs, the Scheduler sends the operation request to the Bus concurrency controller (BCC). The BCC checks the status of the operation and forwards it to the appropriate node in the cluster. This forwarding is done via the BNO manager.

Once the operation request is sent to the appropriate node, the transaction is put on hold until the result arrives to the BNI manager. The manager forwards the result back to BCC which forwards it back to the Scheduler.

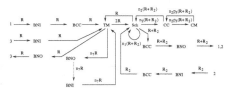

R: Incoming create request arrival rate to a new transaction

R_2: Incoming bus-level operation request arrival rate to an existing transaction

n_1: Number of CPU-operation rounds in a bus-level operation

n_2: Number of CC-operation rounds in a bus-level operation

p_3: Probability for a CC-operation to need CM-services

n_7: Number of slave bus-level commit protocol message rounds

Figure 5.18: Slave bus-level transaction pattern.

When the transaction is finished, the TM initiates a global commit procedure to all its sub-transactions in slave nodes. It needs n_5 rounds to the slave nodes to get a result. The average number of slave nodes here is n_6. Thus the total number of messages in the bus-level commit procedure is $n_5 n_6 R / n_0$.

Once the global commit procedure is finished, the transaction is either committed or aborted. In both cases the result is sent back to the cluster entry node via the BNO manager. The result is forwarded back to the request source like in the previous pattern.

Slave bus-level transaction pattern

The third transaction pattern describes a slave transaction in a bus-level distributed transaction. It can be seen in Figure 5.18. This pattern occurs whenever a bus-level master transaction wants to create a slave transaction to the analyzed node. It happens both after a master bus-level transaction pattern and also after a master or slave cluster-level transaction pattern since all three transaction types may be master transactions in a bus-level transaction.

The slave bus-level transaction pattern has three entry points: when a slave transaction is created, when a request arrives to an existing slave, and when the master initiates a global commit procedure.

First the node receives a request of a master from another node. The requests arrive at rate R. In the node the request arrives to the BNI manager which forwards it to the BCC of the node. The BCC checks if a slave transaction has already been created. Since this is the first call to the BCC, a slave transaction has not been created yet. The request is sent to the TM for slave transaction creation.

The TM creates a new transaction which executes the request. It needs to access local data and perhaps do CPU operations. The total number of local data access operations is n_1 and CPU operations is n_2. When the first request has been calculated, the result is sent to the BCC which returns it back to the requesting BCC via the BNO manager. After that the transaction is put into a wait state until the next request arrives.

When the master transaction sends a later request, the BCC already knows to which transaction in this node to forward it. The request is sent via BNI to the BCC which forwards it to the Scheduler. After that the operation needs n_1 CPU rounds, n_2 CC rounds, and $n_2 p_3$ CM rounds. After data delivery and possible CPU operations the result of the request is sent back to the master transaction via BCC. The arrival rate of this type of requests is R_2.

Once the master is ready to commit, it sends a commit request via BNI to the TM. The slave needs to take n_7 rounds before the commit result of the transaction is known. After that the transaction is either committed or aborted. In both cases the slave transaction sends an acknowledgment message back to the master transaction. The acknowledgment message goes from the TM via the BNO to the master transaction node. The master transaction is responsible for sending the final transaction result back to the requester.

In this pattern we thus have two arrival rates: R for new slave transaction creation requests, and R_2 for old slave transaction operation requests. These arrival rates have a relationship that depends on the master transaction parameters on master bus-level transaction pattern, and master and slave cluster-level transaction patterns. It is simpler to use two request arrival rates and later verify how they are related. We will analyze the relationship in Section 5.4.1.

Master cluster-level transaction pattern

The fourth transaction pattern describes a distributed transaction between several clusters. The pattern can be seen in Figure 5.19. In the pattern we see the master transaction.

The incoming request arrival rate is R. When a request arrives to Net In, it is first checked to see whether it is an X.500 request. In such a case it is sent to the X.500 manager for interpretation. This happens with probability p_9. The X.500 needs n_9 rounds to CDM for interpretation. The CDM may need to access a disk manager DM on some of the rounds. The probability for a disk access is p_{10}.

When the interpretation n_9 rounds are over, the interpreted request is sent to the GTM. If the request is not an X.500 request, it is directly sent to the GTM.

The GTM decides which parallel database node receives the request. The node N_i request arrival rate is R/n_0 since we assume that the requests are distributed evenly on all parallel database nodes.

The TM creates a transaction that has four types of operations: n_1 CPU operations, n_2 local database operations, n_4 bus-level operations, and n_8 network-level operations.

The CPU, local database, and bus-level operations follow the same pattern as in the previous case. A network level request is first forwarded to the BCC. It sends the request via the BNO manager to the cluster entry node. There the cluster entry node Bus In manager forwards it to the Global concurrency controller (GCC). The number of global operations from this node is $n_8 R/n_0$.

The GCC sends the request via the Net Out manager to the appropriate cluster. Once the result is received from the appropriate cluster to the Net In manager, it is sent via the GCC and the Bus Out manager to the correct node. There the node BNI manager sends the result to BCC which forwards it to the Scheduler. When the result is received from the other cluster, the master transaction may proceed.

Once the transaction is ready to commit, the commit is done at two levels. First, at bus-level the master transaction follows a bus-level commit procedure to decide if the transaction can commit at the cluster level. If it can, the result is sent back to the GTM which executes a

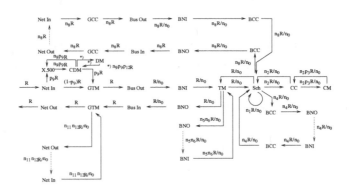

R:	Incoming request arrival rate
n_0:	Number of nodes in a cluster
n_1:	Number of CPU-operation rounds
n_2:	Number of CC-operation rounds
p_3:	Probability for a CC-operation to need CM-services
n_4:	Number of bus-level BCC operation rounds
n_5:	Number of master bus-level commit protocol message rounds
n_6:	Average number of slave nodes in a bus transaction
n_8:	Number of cluster-level BCC operation rounds
n_9:	Number of X.500-CDM-X.500 rounds
p_9:	Probability for an incoming request to be of X.500 type
p_{10}:	Probability for CDM to need DM services
n_{11}:	Number of master cluster-level commit protocol message rounds
n_{12}:	Average number of slave clusters in a cluster transaction

Figure 5.19: Master cluster-level transaction pattern.

network-level commit procedure to decide if the transaction can commit globally. At bus level n_5 rounds are needed for completing the atomic commit procedure. The number of slave nodes to participate in the procedure is n_6. Hence $n_5 n_6$ messages are sent and received in the protocol.

At cluster level n_{11} rounds are needed, and the average number of slave clusters in a cluster-level transaction is n_{12}. Hence $n_{11} n_{12}$ messages are sent and received in the protocol. Once the GTM has decided the result for the committing transaction (commit or abort), the result is sent back to the requester via the Net Out manager.

Slave cluster-level transaction pattern

The fifth transaction pattern describes a transaction that is a slave in a network-level distributed transaction. The pattern is in Figure 5.20. Like the slave bus-level transaction pattern, this pattern also has three entry points: a new transaction creation, a request to an existing transaction, and a global cluster-level commit protocol request.

First, a new operation request arrives from a master cluster. The arrival rate of these requests is R. The Net In manager receives it and forwards it to the GCC. The GCC notifies that the request is from a master transaction that does not yet have a slave in this cluster. It sends the request to the GTM which creates a new cluster-level slave transaction.

The freshly created transaction is sent via Bus Out to an appropriate node in the cluster. The node BNI manager receives the transaction and forwards it to the TM in the node. The TM creates a new bus-level transaction and starts execution. The number of incoming requests to a parallel database node is R/n_0.

The new transaction executes the operations that were received from the master transaction. The transaction execution may trigger bus-level distributed operations which make the slave cluster transaction a master in a bus-level transaction. This part of the analyzed transaction pattern has been described in the master bus level transaction pattern analysis.

Once the current operation execution is over, the results are sent via the BCC out to the GCC that returns the result to the GCC of the master transaction cluster. The bus transaction itself is idle waiting for the next requests to arrive.

The next time an operation request arrives to this slave transaction, the GCC recognizes it

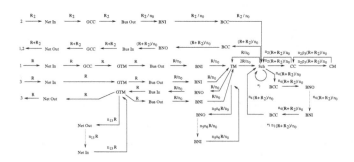

R: Incoming create request arrival rate to a new transaction

R_2: Incoming cluster-level operation request arrival rate to an existing transaction

n_0: Number of nodes in a cluster

n_1: Number of CPU-operation rounds in a cluster-level operation

n_2: Number of CC-operation rounds in a cluster-level operation

p_3: Probability for a CC-operation to need CM-services

n_4: Number of master bus-level operation rounds in a cluster-level operation

n_5: Number of master bus-level commit protocol message rounds

n_6: Average number of slave nodes in a bus transaction

n_{13}: Number of slave cluster-level commit protocol message rounds

Figure 5.20: Slave cluster-level transaction pattern.

and sends it directly via the bus to the appropriate node. The Scheduler activates the sleeping transaction which then executes the received operations. Again the operations may be to bus-level slave transactions as well. When the operations have been executed the results are sent back to the master cluster-level transaction the same way as when the transaction was first created.

When the master cluster transaction is ready to commit, it sends a prepare message to the slave transaction. The prepare message is sent to the GTM which sends it to the bus transaction at the correct node.

If the slave transaction is a master at a bus level, it executes a bus master commit procedure pattern that we described in Section 5.4.1. Once the bus-level voting is finished, the slave transaction knows how it will vote at a cluster-level voting procedure. Its vote is sent to the GTM which will participate in a cluster-level atomic commit procedure. The cluster-level vote needs n_{13} rounds to complete.

Once the cluster-level commit procedure is finished, the result is sent back to the node. The node TM activates the bus transaction which then finishes the bus-level atomic commit procedure (where it is a master transaction). Finally, the slave cluster-level transaction sends an acknowledgment to the master transaction. After that the slave transaction commits or aborts itself depending on the result of the cluster-level votes.

X.500 service reply/directory update

The last transaction pattern describes X.500 cluster replies and directory updates. They both follow the same pattern that can be seen in Figure 5.21. This is the simplest pattern. The other X.500 operation, namely X.500 service request with interpretation is part of the master cluster transaction pattern.

The Net In manager receives a request and forwards it to the X.500 manager. The X.500 manager does an interpretation and sends the result to the CDM manager. The CDM does necessary updates and returns acknowledgment to the X.500. After that the result is returned via Net Out back to the client. The CDM may need to consult the DM with probability p_{10}.

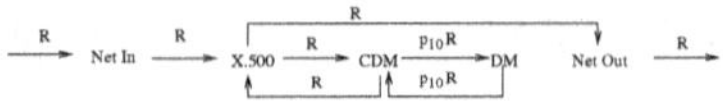

R:　　Incoming request arrival rate

p_{10}:　Probability for CDM to need DM services

Figure 5.21: X.500 service reply/directory update pattern.

Relationships between master and slave transactions

Both cluster-level and bus-level distributed transactions have master and slave transaction patterns. The master transactions initiate all actions. The slaves respond to master requests. Thus, we have a clear relationship between master and slave transaction patterns.

We will first analyze the simpler case: the relationship between cluster-level master and slave transaction patterns. This is a simpler case than the bus-level case since here we have only two patterns to deal with. In the bus-level case we have four patterns to consider since also cluster-level master and slave transactions may participate as masters in bus-level distributed transactions.

In the cluster-level transaction pattern, from the slave point of view we have an arrival rate of R incoming create requests and an arrival rate of R_2 incoming regular operations. These requests arrive from all other $c_c - 1$ clusters. Thus, the number of arriving requests from a single master cluster is $R/(c_c - 1)$ for create requests, and $R_2/(c_c - 1)$ for incoming regular operations.

From a master transaction point of view the total arrival rate of distributed operations is $n_8 R_M$ where n_8 is the average number of distributed operation rounds, both create requests and regular operations; and R_M is the arrival rate of all requests in a master node. Of the n_8 rounds, one round is for a new slave transaction creation and $n_8 - 1$ rounds are for regular operations. These operations are divided to all n_{11} slave nodes that participate in a distributed transaction. Thus, we have the following relationships:

$$R\frac{1}{c_c - 1} = R_M\frac{1}{n_{11}}, \text{ and } R_2\frac{1}{c_c - 1} = R_M\frac{n_8 - 1}{n_{11}}. \tag{5.46}$$

This gives us the relationships to R and R_2 in the slave cluster-level transaction pattern:

$$R = R_M \frac{c_c - 1}{n_{11}}, \text{ and } R_2 = R_M \frac{(n_8 - 1)(c_c - 1)}{n_{11}}. \tag{5.47}$$

Also a relationship between the number of master cluster commit protocol rounds n_{12} and slave cluster commit protocol rounds n_{13} exists in the slave cluster-level transaction pattern. For every sent request to the network there must be someone to receive it, and vice versa. In the analysis we use a model where the master initiates all transactions between slaves and slaves respond to the master only. This gives us a relationship $n_{13} = n_{12} - 1$ because on each request the slave will respond, and the first prepare round is received at route Net In $\rightarrow$GTM$\rightarrow$Bus Out$\rightarrow$ etc. so that the bus-level commit procedure can be initiated and finished as well.

The bus-level master and slave transaction relationship is similar to the cluster-level relationship. However, in the bus-level case also cluster master and slave transactions may be bus-level master transactions. These requests must be included in the analysis.

From the slave point of view the situation is similar to the cluster-level slave transactions. We have an arrival rate R for incoming create requests, and an arrival rate R_2 for incoming regular operations. The total number of nodes in a cluster is n_0, and excluding our slave node the rest may send create requests to us. Thus, the total arrival rate of create requests from a single node is $R/(n_0 - 1)$, and the total arrival rate of regular operations is $R_2/(n_0 - 1)$.

From the master point of view, we have three cases to consider: master bus-level transactions without cluster-level distribution, master bus-level transactions that are cluster-level masters, and master bus-level transactions that are cluster-level slaves. We analyze these in turn.

The master bus-level transaction has arrival rate R_m, and the number of bus-level requests to its slaves is n_4. This includes both creation requests and regular operation requests. We rename the n_4 constant to n_{4m} to avoid confusion with other local n_4 constants later.

The master cluster-level transaction has similar figures: R_M arrival rate and an average n_{4M} bus-level operations in a master bus-level transaction. Similarly the slave cluster-level transaction has $R_S + R_{2S}$ request arrival rate and an average n_{4S} bus-level operations in a master bus-level transaction.

With these values, and with the analysis in the cluster-level transaction case, we get the following formulas for slave bus-level transaction pattern:

$$R = R_c \frac{n_0 - 1}{n_6}, \text{ and } R_2 = R_o \frac{n_0 - 1}{n_6}, \tag{5.48}$$

where

$$R_c = R_m + R_M + (R_S + R_{2S}), \tag{5.49}$$

and

$$R_o = (n_{4m} - 1)R_m + (n_{4M} - 1)R_M + (n_{4S} - 1)(R_S + R_{2S}). \tag{5.50}$$

Just like in the cluster-level commit routine, also in the bus-level there is a relationship between the number of master bus-level commit protocol rounds n_5 and slave bus-level commit protocol rounds n_7. For every sent request to the bus there must be someone to receive it, and vice versa. In the analysis we use a model where the master initiates all transactions between slaves and slaves respond to the master only. This gives us a relationship $n_7 = n_5 - 1$ because on each request the slave will respond, and the first prepare round is received at route BNI $\rightarrow$TM$\rightarrow$ etc. so that the bus-level commit procedure will be initiated.

5.4.2   Manager clusters and service times

In the Cluster entry node, the clearly hardware-driven managers are Net In, Net Out, Bus In, and Bus Out. Each of these receives requests and forwards them to appropriate software managers, to the network, or to the bus. We consider them to form a manager cluster although basically we could consider each of them to form a cluster on their own.

The software managers in the cluster entry node are X.500, CDM, GTM, GCC, and LCM. They clearly form a cluster since they share the same CPU.

The DM is a disk controller. As such it is neither a network hardware entity nor a software manager. It forms a cluster alone in the model.

Second, we will consider manager service times. In principle every transaction pattern may have local service times since the service times are related to incoming request flows. Yet most managers have equal service times regardless of the transaction pattern since similar incoming flows on each pattern present similar actions.

The service times depend on the hardware used, such as CPU, memory, and disk features. In the analysis we assume that we have state-of-the-art personal computer hardware. We are not interested in analyzing architectures that are too expensive to build or use.

We consider the CPU and simple hardware service times to vary from 0.01ms to 1ms. With a state-of-the-art CPU 1ms is over 500,000 machine operations when one operation takes four CPU cycles.

The managers in the cluster-node analysis are both in the cluster entry node and in the database nodes. The cluster entry node managers are Net In, Net Out, X.500, CDM, DM, GTM, GCC, LCM, Bus In, and Bus Out. The database node managers are BNI, BNO, TM, Scheduler, CC, CM, and BCC. Next to these managers, the Network and the Bus are modeled as managers. We now consider here the service execution times of all managers in turn.

The Net In manager accepts requests from the network and forwards them to appropriate cluster entry node managers. The operations are straightforward and simple to implement. We consider the manager to take 0.01ms. Similarly the Net Out manager receives requests from various sources and forwards them to the network. We consider this manager to need 0.01ms for service execution as well.

The X.500 manager is responsible for translating incoming X.500 requests to the Yalin internal database model. This is not a straightforward task. In the operation the X.500 manager may need several consultation rounds to the CDM. We consider each X.500 round to take 1ms.

The CDM holds information of the Yalin data tree. This information is needed for X.500 operations. The manager holds information in the main memory and transfers it from there via directories. We expect a single access to take 0.5ms.

Sometimes the CDM must consult the DM that holds directory information on a disk. The DM is a regular hardware manager for disks. Each disk operation has elements that are equal to the earlier Database entry node analysis. Thus we can use the same parameter values that we used in the global entry disk manager analysis. This gives us the disk access time

$$t_{\text{disk}} \approx 12.3\text{ms.} \tag{5.51}$$

As earlier, the average service time is equivalent to the service time of a good modern hard disk.

The GTM creates new global transactions and manages global commit procedures. These tasks are not straightforward, so we consider them to take 1ms each. The same is true for the GCC manager which must decide which cluster to forward a service request. We consider this to take 1ms on each request as well.

The LCM is an entry point for local transactions. It both receives requests from Net In and decides which database node in the cluster should receive the request, and receives responses from local database transactions and forwards them to the requesters. These are relatively straightforward tasks since the receiver is already known when LCM services are called. We consider this to take 0.1ms.

The Bus In manager receives data from the common bus that is sent from database nodes to the cluster entry node. The Bus Out manager sends cluster entry node data to the common bus for database nodes. Thus these managers are similar to Net In and Net Out. We consider also these managers to take 0.01ms for each service request.

On each database node, the TM creates new transactions and participates in bus-level commit procedures. These tasks are not straightforward so we consider them to take 1ms each.

A Scheduler in a database node has two tasks: it either forwards a data item request to CC or BCC, or it executes a CPU operation. The former we consider to take 1ms, the latter 5ms.

The CC and BCC manage operations for local and external data items, respectively. They do not have access to the physical database but they may access data items that are in the main memory. The CC manages local data items. We consider this task to take 1ms for each data item. The BCC decides where an external request is forwarded. This information is in local directories that are mostly in the main memory. We consider also this operation to take 1ms.

The CM is a disk manager like the DM in the cluster entry node. We use the same parameters to the CM that we used for the DM. Thus, the average service time is about 12.3ms. Similarly the database node managers BNI and BNO are equal to cluster entry node managers Bus In and Bus Out. We consider BNI and BNO both to take 0.01ms on each request.

The network and bus are also considered managers in the model. Both of them have service times of types $t =$ packet size$/$speed where the packet size is the average size of the sent or received packet in bits, and speed is the network or bus speed in bits/second. We consider the

bus	Bus speed bits/second	25Mb/s
network	Network speed bits/second	64kb/s
busblock	Bus block size	256 B
netblock	Net block size	256 B

Table 5.2: Bus and network parameters.

average packet size to be 256 bytes since most traffic both in network and in the bus is very small. The network speed is 64kb/s and bus speed is 25Mb/s. With these values the network service time is 32ms and the bus service time is 0.08ms. However, we very soon found out that the value 64kb/s gives boring results. The network overloads immediately and makes all global transactions impractical. Due to this we used a higher network speed 1Mb/s.

In the analysis we will examine the bus and network, cluster entry node managers, and database node managers. We use the average service execution times and service analysis parameters as described earlier. The service times and parameters are also summarized here.

Table 5.2 summarizes bus and network parameters. The network and bus speeds are based on current hardware, except that the network speed is still relatively low. Yet when we want to analyze a distributed database it is better to be conservative with network estimations. Although Yalin is not intended to work in a public network, network capacity can still become a bottleneck. The network speed value 1Mb/s should be realistic enough while not giving too optimistic estimates.

Table 5.3 summarizes the manager service times. The service times have been described earlier and need not be repeated here. It should be noted that the CEN disk manager and the node CM both use disk values that are derived from known hard disk parameters.

Excluding the Scheduler, each manager has a single service time for all incoming request flows. This is a reasonable assumption since the managers mostly have few tasks to manage. An exception to this rule are transaction managers. They process transaction creation, deletion, and global commit procedure tasks. However, since all these tasks are about equally complex, we have taken an approach to use a single service time for all the tasks. The Scheduler, how-

t_1	CEN Net In service time	0.000010s
t_2	CEN Net Out service time	0.000010s
t_3	CEN X.500 interpreter service time	0.001000s
t_4	CEN Cluster directory manager service time	0.000500s
t_5	CEN Disk manager service time	0.012331s
t_6	CEN Global transaction manager service time	0.001000s
t_7	CEN Global concurrency controller service time	0.001000s
t_8	CEN Local call manager service time	0.000100s
t_9	CEN Bus In service time	0.000010s
t_{10}	CEN Bus Out service time	0.000010s
t_{11}	Node Bus-in service time	0.000010s
t_{12}	Node Bus-out service time	0.000010s
t_{13}	Node Transaction manager service time	0.001000s
t_{14a}	Node Scheduler quick service time	0.001000s
t_{14b}	Node Scheduler slow service time	0.005000s
t_{15}	Node Concurrency controller service time	0.001000s
t_{16}	Node Cache manager service time	0.012331s
t_{17}	Node Bus concurrency controller service time	0.005000s
t_{18}	Network service time (1Mb/s speed)	0.002000s
t_{19}	Bus service time (25Mb/s speed)	0.000082s

Table 5.3: Service time default values.

ever, needs two service times: one for forwarding read and write operations to the concurrency controller, and one for executing CPU operations.

5.4.3 Analyzed transactions

The Yalin system has 23 transaction types, as analyzed in Section 4.4. In the bottleneck analysis we are mostly interested in transaction sizes and how they are distributed at cluster and bus levels. For the analysis we need to know transaction lengths, relative transaction frequencies, and probabilities for transaction distribution levels (local, bus-level or cluster-level). Most of this information can be extracted from the IN/GSM data and transaction analysis. We summarize the analysis results in Table 5.4.

The relative transaction frequencies are estimated from IN/GSM transaction type analysis using relative frequency weights. Rare transaction types, such as Query and management transaction types, have a weight value 1. The higher the weight value of a transaction, the more frequent it is compared to the value 1 transaction types. For instance, we estimated transactions of the transaction type Profile to be a thousand times more common than transactions of the transaction type Query. Thus, the weight value of the Profile transaction type is 1000. The actual frequencies depend on request arrival rates and transaction type weights.

Each transaction has a priority that affects the scheduling order. We have ignored the priorities in the analysis since they are not important when estimating Yalin bottlenecks. When resources are sufficient, all transactions will eventually execute. The transaction priorities are directly comparable to transaction frequencies since high-priority transactions are the shortest and most common in Yalin. The bottleneck analysis gives a good estimation of the average service and execution times; the times can then be compared to transaction deadlines.

In the transaction length estimation we can use the information of probable class access as summarized in Table 4.1. On each transaction type the table lists IN/GSM object classes that are accessed by transactions of that type. We convert this information to CPU and CC operations with a simple algorithm. On each read access to an IN/GSM object class, we assume two CC operations: one for reading data and one for directory access. On each write access we assume

Transaction type	Weight	m_1	m_2	n_1	n_2	Local p	Bus p	Clstr p
Location update	10000	0	0.33	0	1	0	0	1
Initialize call	1000	1	3	19	57	0.2	0.4	0.4
Answer call	900	1	3	15	45	0.2	0.4	0.4
Finish call	900	1	3	14	42	0.2	0.4	0.4
Translate	100	1	3	10	30	0.8	0.2	0
Charge	900	1	3	7	21	0.2	0.4	0.4
Mobile	100	1	3	8	24	0	0	1
Activate	500	1	3	8	24	0.8	0.2	0
Profile	1000	1	3	8	24	1	0	0
Create Conf	10	1	3	6	18	0.8	0.2	0
Add member	50	1	3	3	9	0.4	0.3	0.3
Mass write	0	1	3	9	1	1	0	0
Mass read	0	5000	10000	10000	20000	0	0.6	0.4
Account mgmt	1	1	3	8	24	0.4	0.3	0.3
Conference mgmt	1	1	3	6	18	0.8	0.2	0
CUG management	1	1	3	3	9	0.4	0.3	0.3
EIR management	1	1	3	3	9	0.4	0.3	0.3
Phone line mgmt	1	1	3	7	21	0.4	0.3	0.3
Service mgmt	1	1	3	3	9	0.4	0.3	0.3
Subscriber mgmt	1	1	3	9	27	0.4	0.3	0.3
Teleop mgmt	1	1	3	3	9	0.4	0.3	0.3
VPN management	1	1	3	6	18	0.8	0.2	0
Query	1	50	100	1400	2800	0.4	0.3	0.3

Weight:	Relative transaction arrival frequency	m_1:	CPU operation multiplier
m_2:	CC operation multiplier	n_1:	Total number of CPU operations
n_2:	Total number of CC operations	Local p:	Local transaction probability
Bus p:	Bus-level transaction probability	Clstr p::	Cluster-level transaction probability

Table 5.4: Transaction parameters.

three operations: directory, reading and writing. Since the read and write access information is present in the analysis, by adding the assumed operations we get the first estimate of needed CC operations. However, these values indicate a situation where a transaction accesses only one object of each object class. Usually transactions access several objects and do CPU operations to them. Due to this we include two multipliers: m_1 for CPU operations and m_2 for CC operations. The multipliers are used to estimate how many objects a transaction accesses on average and how many CPU operations are needed to calculate the final results. The multiplier values are calculated from the IN/GSM data analysis and from Yalin architecture transaction types and accessed objects.

Next to the transaction sizes, relative transaction frequencies are also estimated. The estimations are based on IN and GSM transaction and object analysis, but nevertheless they are guesswork. Finally for each transaction we have estimated the probability for a local transaction, a bus-level transaction, and a cluster-level transaction. These estimates are based on the transaction types and their descriptions in Section 4.4. The idea in the estimates is that most transactions are local unless we know from the description that the transaction type has a different distribution profile.

The Mass write and Mass read transaction types are excluded from the analysis. The Mass write transaction needs a database node that is tailored for this transaction alone. The Mass read transaction needs mostly access to the Mass write database node. These transaction types would dominate analysis if they were used together with other transaction types.

5.4.4   Bottleneck analysis

The most important question in the Yalin architecture is how suitable it is for an IN/GSM database architecture. The best analysis for this is to see how well various managers can handle transaction arrival rates. In this section we will analyze the Yalin manager bottlenecks, manager service times, and average transaction execution times.

The bottleneck analysis variables are summarized in Table 5.5. The default number of clusters in the analyzed system is 10. This value represents a reasonable-sized distributed database

c_c	Number of clusters in the system	10
n_0	Number of nodes in a cluster	10
n_1	Number of CPU operation rounds	(Table 5.6)
n_2	Number of CC operation rounds	(Table 5.6)
p_3	Probability for CM access from CC	0.4
n_4	Number of bus-level BCC operation rounds	(Table 5.6)
n_5	Number of master bus-level commit protocol message rounds	2
n_6	Average number of slave nodes in a bus transaction	3
n_7	Number of slave bus-level commit protocol message rounds	1
n_8	Number of cluster-level BCC operation rounds	(Table 5.6)
n_9	X.500$\rightarrow$CDM$\rightarrow$X.500 rounds	6
p_9	Probability for incoming request to be of X.500 type	0.01
p_{10}	Probability for CDM to need disk access	0.8
n_{11}	Number of master cluster-level commit protocol message rounds	2
n_{12}	Average number of slave clusters in a cluster transaction	3
n_{13}	Average number of slave cluster-level protocol message rounds	1

Table 5.5: Bottleneck analysis variables and default values.

Pattern	n_1	n_2	n_4	n_8
Local	1.48	5.06	-	-
Master bus	1.14	3.99	1.13	-
Slave bus	0.33	1.00	-	-
Master cluster	0.84	2.52	1.12	1.12
Slave cluster	0.33	1.00	0.00	-
X.500 reply/update	-	-	-	-

Table 5.6: Number of operation rounds in transaction patterns.

as each cluster is also a parallel database with 10 database nodes.

The number of CPU operation rounds, CC operation rounds, BCC operation rounds, and cluster-level BCC operation rounds have been excluded from the table. These values are transaction pattern specific. Their values are summarized in Table 5.6. Local transaction patterns cannot have bus-level or cluster-level operations and bus-level patterns do not have cluster-level operations. The X.500 directory update and service replies do not have any operations since they may access only the X.500 directory.

The Slave cluster transaction pattern could in principle have bus-level master operations. However, their probability in the analysis was so close to zero that we rounded the value to zero. It simplified the analysis a little.

The probability for a CM access from the CC basically defines the relative sizes of the main memory and the disk in a database node. The default value 0.4 states that most active data resides in the main memory while less common data is accessed from a disk. This value is large enough to include necessary updates to the disk so that the major copy of a data item resides there. Hence, Yalin is not a main memory database management system architecture.

The number of master bus-level and master cluster-level commit protocol message rounds are chosen for a normal two-phase commit protocol. The average number of slave nodes in a bus-level and cluster-level transactions is just a guess. The value cannot be very large since the number of distributed operations from a master node is small. That value is calculated from

p_a	Probability for local transaction pattern	0.712833
p_b	Probability for master bus-level transaction pattern	0.094235
p_c	Probability for slave bus-level transaction pattern	N/A
p_d	Probability for master cluster-level transaction pattern	0.092932
p_e	Probability for slave cluster-level transaction pattern	N/A
p_f	Probability for X.500 transaction pattern	0.100000

Table 5.7: Transaction pattern probabilities.

transaction type properties.

The number of X.500 interpretation rounds is also a guess. The value depends on how complex the X.500 operations generally are. The same is true of the probability for incoming X.500 type requests and the probability for CDM disk access.

The default CDM disk access probability is 0.8. This value indicates a small hot spot in the database so that most data item requests must visit the physical database. This is true in the X.500 directory information since the X.500 operations are probably not limited to a specific area of the database.

Table 5.7 summarizes various transaction pattern probabilities. The probabilities for pattern a (local transaction), pattern b (bus-level master transaction), and pattern d (cluster-level master transaction) are calculated from the IN/GSM transaction and data analysis. The values are normalized so that the probability for pattern f (X.500 reply/update transaction pattern) is a free value. The probabilities for pattern c (slave bus-level transaction pattern) and pattern e (slave cluster-level transaction pattern) are calculated from their relationships to the other transaction patterns as described in Section 5.4.1.

Bus and network analysis

In the bus and network analysis we have first used the default values in Table 5.5. The traffic in a bus is between the database nodes and the cluster entry node. The traffic in the network is considered to include all clusters.

In the first case (Figure 5.22) we use the default X.25 signal speed 64kb/s for the network. The results are devastating. The network becomes an immediate bottleneck in the system. This result is not surprising. Even a small number of distributed transactions in the system causes heavy traffic between clusters. If all traffic is considered to use the same wires with 64kb/s speed then the global commit protocols alone cause a severe burden to the network. If the 64kb/s is the network speed limit then distribution is out of the question in Yalin.

Fortunately we are not restricted to the old X.25 network. The third generation mobile networks already use TCP/IP based networks that offer higher network speeds than the X.25. The same can be true with the Yalin architecture. Only the entry points may need access to X.25. The rest of the system, including connections to clusters, may use faster TCP/IP-based connections.

In the rest of the analysis we assume that the network speed is 1024kb/s. With this speed we get more promising results, as can be seen in Figure 5.23. The network can handle about 140 arriving requests/second in a cluster (when we have 10 clusters in the network) and it no longer becomes the limiting factor.

The cluster bus throughput is much better than the network throughput. Although traffic in a cluster node is higher than traffic between two clusters, the bus speed is better. We use a bus speed of 25Mb/s which is a reasonable value for an external bus that connects shared-nothing databases. With this speed, having 10 nodes in a cluster does not create a bus bottleneck. In fact, as much as 150 nodes in a cluster may be supported (Figure 5.24). However, when extremely large numbers of nodes are modeled in the analysis, even small changes in parallel transaction probabilities affect the results. We believe that a more realistic result can be seen in an analysis with 50 nodes and bus block size 512 bytes. The results of the analysis are in Figure 5.25.

The scalability of the Yalin bus is very good. This is basically due to the fact that only 10 % of the requests trigger a bus-level transaction. These values have been calculated from the known transactions, their assumed distribution between local, bus-level, and cluster-level transactions, and from their analyzed database operations. However, the atomic bus block size affects the Yalin shared-nothing architecture throughput. When the block size is raised to 1024 bytes, the system can handle only 10 nodes in a cluster while with 512 byte blocks it can handle

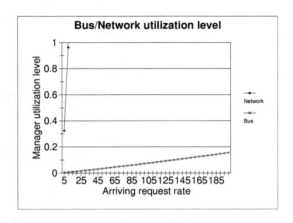

Number of clusters in the system	10
Number of nodes in a cluster	10
Network speed	64kb/s
Network block size	256B
Bus speed	25Mb/s
Bus block size	256B
Local transaction probability	0.712833
Bus-level transaction probability	0.094235
Cluster-level transaction probability	0.092932
X.500 interpretation probability	0.100000
X.500 interpretation rounds	6
CM access probability	0.4

Figure 5.22: Bus and network utilization, network speed = 64kb/s.

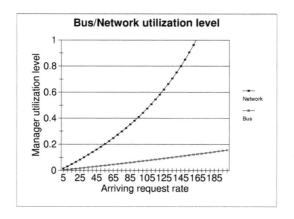

Number of clusters in the system	10
Number of nodes in a cluster	10
Network speed	1Mb/s
Network block size	256B
Bus speed	25Mb/s
Bus block size	256B
Local transaction probability	0.712833
Bus-level transaction probability	0.094235
Cluster-level transaction probability	0.092932
X.500 interpretation probability	0.100000
X.500 interpretation rounds	6
CM access probability	0.4

Figure 5.23: Bus and network utilization, network speed = 1024kb/s.

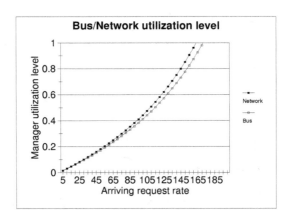

Number of clusters in the system	10
Number of nodes in a cluster	150
Network speed	1Mb/s
Network block size	256B
Bus speed	25Mb/s
Bus block size	256B
Local transaction probability	0.712833
Bus-level transaction probability	0.094235
Cluster-level transaction probability	0.092932
X.500 interpretation probability	0.100000
X.500 interpretation rounds	6
CM access probability	0.4

Figure 5.24: Bus and network utilization, 150 nodes.

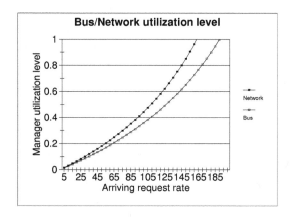

Number of clusters in the system	10
Number of nodes in a cluster	50
Network speed	1Mb/s
Network block size	256B
Bus speed	25Mb/s
Bus block size	512B
Local transaction probability	0.712833
Bus-level transaction probability	0.094235
Cluster-level transaction probability	0.092932
X.500 interpretation probability	0.100000
X.500 interpretation rounds	6
CM access probability	0.4

Figure 5.25: Bus and network utilization, 50 nodes, bus block size = 512 bytes.

50 nodes. This can be seen in Figure 5.26.

Since the previous results of the bus scalability sounded almost too good, we wanted to try the default values and an artificial probability of 0.8 for bus-level transactions (compared to the calculated probability 0.09). Even then the bus behaved well enough as can be seen in Figure 5.27. In the figure we have 10 database nodes and atomic blocks of size 256 bytes. The bus never becomes a bottleneck with these values.

Naturally these values are only estimates based on the IN/GSM object class and transaction analysis. Yet real system object classes and transactions should not be too far from the Yalin system. Parallelism really helps in Yalin since most transactions are local to a single node. Local transactions can be distributed to several database nodes in a cluster with little extra burden from bus-level commit procedures for bus-level transactions.

As a conclusion to the bus and network analysis we can state that parallel database nodes are a better alternative than raw distribution in Yalin. It may even be possible to use a single IN/GSM parallel database management system and hence avoid distribution completely. On the other hand, since most transactions are not distributed, geographically distributed clusters shorten access times for local transactions since the network delay from a client to the database becomes shorter.

It should also be noted that we did not analyze specialized nodes that are related to mass calling and televoting. When using specialized clusters and nodes, those services can be simplified. This is an especially suitable idea for Yalin clusters. One database node in a cluster may be devoted to mass calling and televoting service alone. Mass calling writes are forwarded directly to the node. Mass calling reads are implemented similarly to other services, but they are executed in the mass calling database node and then perhaps use bus-level subtransactions in other nodes.

Cluster entry node analysis

The Cluster entry node (CEN) has three tasks. It is an entry node for bus-level and local transactions, a communication node for cluster-level transactions, and an interpreter for X.500-based queries and responses. As such it is the only entry to a cluster.

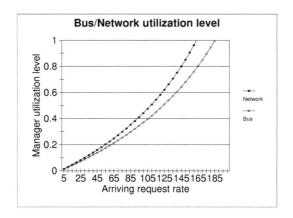

Number of clusters in the system	10
Number of nodes in a cluster	10
Network speed	1Mb/s
Network block size	256B
Bus speed	25Mb/s
Bus block size	1024B
Local transaction probability	0.712833
Bus-level transaction probability	0.094235
Cluster-level transaction probability	0.092932
X.500 interpretation probability	0.100000
X.500 interpretation rounds	6
CM access probability	0.4

Figure 5.26: Bus and network utilization, 10 nodes, bus block size = 1024 bytes.

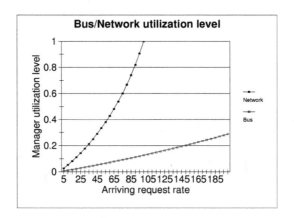

Number of clusters in the system	10
Number of nodes in a cluster	10
Network speed	1Mb/s
Network block size	256 B
Bus speed	25Mb/s
Bus block size	256 B
Local transaction probability	0.157399
Bus-level transaction probability	0.800000
Cluster-level transaction probability	0.020520
X.500 interpretation probability	0.022081
X.500 interpretation rounds	6
CM access probability	0.4

Figure 5.27: Bus and network utilization, bus-level transaction probability = 0.8.

In principle a cluster should not need any database nodes since the CEN has a disk manager for X.500 directory information. This manager could be used for regular data management as well. However, in Yalin all transactions are forwarded to database nodes. This simplifies the CEN architecture since transaction scheduling and concurrency control are processed in nodes.

The results of CEN analysis using default values is in Figure 5.28. The values seem to be sufficient for Yalin use. The Global transaction manager (GTM) is the first manager to overload but even it can handle almost 200 requests in a second. The CPU utilization does not become an issue, nor does the Disk manager. In fact, the Disk manager is underloaded. With these values it would be possible to have a hot data database in CEN.

One thing more to notice here is the Local control manager (LCM). It is responsible for collecting local and bus-level requests and forwarding them to the appropriate database nodes. It is not at a risk of an overload at all. The conclusion of this is that it is possible to centralize request collection into the LCM and hence have a single entry to each cluster.

The CEN utilization changes dramatically when the system has 20 clusters instead of the default 10. This can be seen in Figure 5.29. Now the GTM soon becomes a bottleneck. This is due to the extra global commit procedure processing. Similarly the CPU and the GCC can no longer handle high request arrival rates. The LCM overloads as well, but not because of extra local and bus-level requests, but because it has to wait for the CPU.

In Figure 5.30 we have 10 clusters and no cluster-level commit protocols. Now none of the managers become a bottleneck. This also ensures that the bottleneck in the CEN is in the global commit procedure management.

While giving up the global commit protocols may sound promising, the problems soon become unbearable. Without a global commit protocol, transactions may interfere with each other. This easily violates database consistency, for instance when two global transactions want to update the same data item. If some other method is used, such as a correcting transaction, the same problems with request throughput are evident as with global commit protocols, but without the benefits of the protocol.

Since the DM is underloaded even with a high probability of 0.8, an interesting question is to see how much X.500 interpretations Yalin can handle. We analyzed this with the X.500

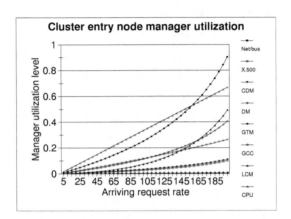

Number of clusters in the system	10
Number of nodes in a cluster	10
Network speed	1Mb/s
Network block size	256 B
Bus speed	25Mb/s
Bus block size	256 B
Local transaction probability	0.712833
Bus-level transaction probability	0.094235
Cluster-level transaction probability	0.092932
X.500 interpretation probability	0.100000
X.500 interpretation rounds	6
CM access probability	0.4

Figure 5.28: Cluster entry node utilization, default values.

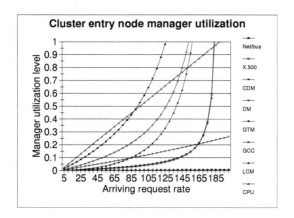

Number of clusters in the system	20
Number of nodes in a cluster	10
Network speed	1Mb/s
Network block size	256 B
Bus speed	25Mb/s
Bus block size	256 B
Local transaction probability	0.712833
Bus-level transaction probability	0.094235
Cluster-level transaction probability	0.092932
X.500 interpretation probability	0.100000
X.500 interpretation rounds	6
CM access probability	0.4

Figure 5.29: Cluster entry node utilization, 20 clusters.

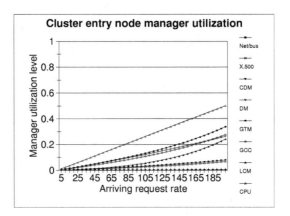

Number of clusters in the system	10
Number of nodes in a cluster	10
Network speed	1Mb/s
Network block size	256 B
Bus speed	25Mb/s
Bus block size	256 B
Local transaction probability	0.712833
Bus-level transaction probability	0.094235
Cluster-level transaction probability	0.092932
X.500 interpretation probability	0.100000
X.500 interpretation rounds	6
CM access probability	0.4

Figure 5.30: Cluster entry node utilization, no cluster-level commit protocol.

interpretation probability value 0.4 (Figure 5.31). The chosen value is very high. Basically it states that requests are not optimized for Yalin at all. Now the DM is the first manager to overload, but still it can handle 150 requests per second.

If the DM overloads first only when the disk access probability is 0.8 and X.500 interpretation probability is 0.4 then the DM definitely is underloaded. Hence, an easy optimization might be to add hot data to the DM and create very simple transactions for such access.

When X.500 is calculated to need 200 interpretation rounds, the cluster entry node utilization is similar to the utilization in the X.500 high probability case (Figure 5.32). Also now the DM is the first manager to overload, but it can still handle about 150 requests in a second. The X.500 and CDM managers can also handle the high interpretation rounds without problems. The reason for this is the low probability for X.500 requests. A complete disaster occurs when the X.500 interpretation probability is high and each interpretation needs a high number of CDM access rounds. This can be seen in Figure 5.33 where the probability for X.500 access is 0.4, and the number of interpretation rounds is as high as 50.

As a summary of the cluster entry node bottlenecks, the node can handle about 150 requests in a second which should be sufficient. The total number of requests to all clusters is naturally 1500 requests in a second which is quite a lot. Yet the cluster entry node disk manager was not even close to overload. It might be a good idea to let some transactions execute in the CEN and that way use the extra disk manager power there. It depends on how much disk resources the X.500 information maintenance takes.

Database node analysis

Each cluster has a Cluster entry node and a set of database nodes. A database node is a member of a parallel database architecture (the cluster) and also a member of a distributed database architecture (the Yalin architecture). All actual data processing is handled in transactions that execute in the database nodes.

The transactions in the analysis consist of three types of operations: data reads, data writes, and CPU operations. Data reads and data writes are processed in the concurrency controller. CPU operations are processed in the Scheduler manager space. All transactions are created in a

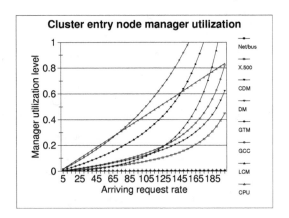

Number of clusters in the system	10
Number of nodes in a cluster	10
Network speed	1Mb/s
Network block size	256 B
Bus speed	25Mb/s
Bus block size	256 B
Local transaction probability	0.47522
Bus-level transaction probability	0.062823
Cluster-level transaction probability	0.061955
X.500 interpretation probability	0.400000
X.500 interpretation rounds	6
CM access probability	0.4

Figure 5.31: Cluster entry node utilization, X.500 interpretation probability = 0.4.

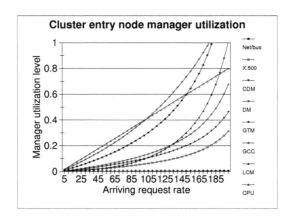

Number of clusters in the system	10
Number of nodes in a cluster	10
Network speed	1Mb/s
Network block size	256 B
Bus speed	25Mb/s
Bus block size	256 B
Local transaction probability	0.712833
Bus-level transaction probability	0.094235
Cluster-level transaction probability	0.092932
X.500 interpretation probability	0.100000
X.500 interpretation rounds	200
CM access probability	0.4

Figure 5.32: Cluster entry node utilization, X.500 interpretation rounds = 200.

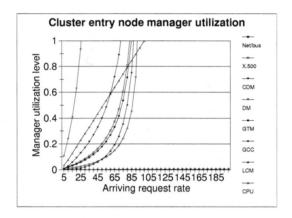

Number of clusters in the system	10
Number of nodes in a cluster	10
Network speed	1Mb/s
Network block size	256 B
Bus speed	25Mb/s
Bus block size	256 B
Local transaction probability	0.475222
Bus-level transaction probability	0.062823
Cluster-level transaction probability	0.061955
X.500 interpretation probability	0.400000
X.500 interpretation rounds	50
CM access probability	0.4

Figure 5.33: Cluster entry node utilization, X.500 interpretation rounds = 50, X.500 interpretation probability = 0.4.

Transaction Manager, executed, and finally committed or aborted. In both the commit and abort cases the result of the transaction is returned to the CEN.

Since both commits and aborts are managed the same way, we do not distinguish them in the analysis. The reason for this is in the nature of real-time transactions. The tight deadlines basically state that it is not reasonable to re-execute a transaction when it is aborted. It would not meet its deadline anyway. We can see this in the next section where we analyze average execution times for various transaction types.

In Figure 5.34 default values are used in the analysis. The scheduler is the first manager to overload but actually all managers overload basically at the same arrival rate of 150 requests per second. The same is true for the CPU and the CM which are hardware managers. Neither of them is a bottleneck.

The database node manager utilization changes dramatically when we calculate the values with 20 database nodes (Figure 5.35). Now none of the managers are even close to overload at 200 requests per second. This implies that Yalin benefits highly from parallel database nodes. This is due to the nature of transactions in nodes. Most Yalin transactions are either local to a single node or do not need much bus-level operations. With these transaction features the nodes really benefit from parallel execution. If the bus-level transaction request arrival rate is high, the Transaction manager would overload. Probably the CPU would overload as well, which would affect the other software managers.

From the analysis we can see that even 200 requests per second is not close to overloading in any of the managers. The scalability of Yalin architecture is good. Since requests are divided equally to all nodes in the cluster, the number of requests per second to one is significantly lower than with the default values. Also the average number of nodes that participate in a bus-level transaction is not changed so less bus-level transaction requests arrive to each database node. as well. Hence, database nodes should not be the first ones to overload when more nodes are added. The bus, as we have seen, is a more realistic candidate for overloading.

The number of nodes that participate in a bus-level transaction affects manager throughput. If we used very high values in the analysis, the transaction manager would overload. Yet this is not an issue. Even if we add nodes, the nature of bus-level transactions does not change. In the

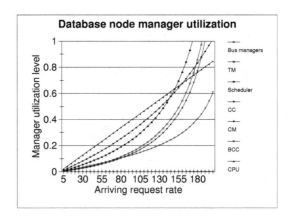

Number of clusters in the system	10
Number of nodes in a cluster	10
Network speed	1Mb/s
Network block size	256 B
Bus speed	25Mb/s
Bus block size	256 B
Local transaction probability	0.712833
Bus-level transaction probability	0.094235
Cluster-level transaction probability	0.092932
X.500 interpretation probability	0.100000
X.500 interpretation rounds	6
CM access probability	0.4

Figure 5.34: Database node utilization, default values.

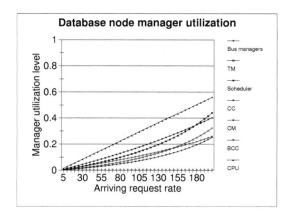

Number of clusters in the system	10
Number of nodes in a cluster	20
Network speed	1Mb/s
Network block size	256 B
Bus speed	25Mb/s
Bus block size	256 B
Local transaction probability	0.712833
Bus-level transaction probability	0.094235
Cluster-level transaction probability	0.092932
X.500 interpretation probability	0.100000
X.500 interpretation rounds	6
CM access probability	0.4

Figure 5.35: Database node utilization, 20 database nodes in a cluster.

analysis we have calculated how much bus-level operations happen on average in a bus-level transaction. This number is relatively small so it is safe to assume that the average number of nodes that participate in a bus-level transaction is not very high.

The CM access probability defines the relative size of the main memory compared to the disk. The used value 0.4 is low which indicates a large main memory. In Figure 5.36 we use a CM access probability of 0.8 which is not that uncommon either. Now the CM is the first manager to overload. As expected with these values, the disk becomes an immediate bottleneck.

The CM utilization can be lowered by adding new database nodes to a cluster. This can be seen in Figure 5.37. While this helps, it still lets the CM dominate the manager utilizations. In optimal settings all managers overload about at the same arriving request level, so none of them becomes a bottleneck. Hence, a better alternative is to add more memory into database nodes and that way lower the disk access probability.

The original value 0.4 for disk access is close to the lowest possible value in a traditional disk-based system. All regularly read-accessed data must reside in a main memory so that read-access is done mostly in the memory. Mostly write-accessed data may reside in a disk as well since changed data must be written to the disk anyway. A pure main memory database would lower the disk access probability but even then change log values must be written to a disk. The Yalin analysis shows that we can use a traditional disk-based approach when we use large enough buffers.

A possible, while questionable, optimization is to exclude the global bus-level commit protocol from bus-level transaction commits. This would violate transaction atomicity and possibly internal and external consistency. The results of this optimization can be seen in Figure 5.38. Compared to the analysis with default values (Figure 5.34), the effect to manager utilizations is surprisingly small. The reason for this is that the number of bus-level transactions is small in the first case, and the bus does not slow down commit execution that much. This is totally opposite to the cluster-level commit procedure where the commit protocol dominated manager utilizations. At the bus-level this kind of an optimization is useless in Yalin. If more transaction throughput is needed, the best way to do it is to add more nodes into a cluster.

The previous analysis is based on the expected transaction distribution between local, bus,

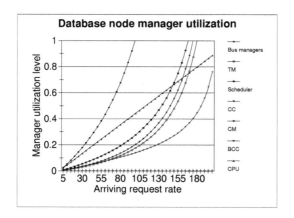

Number of clusters in the system	10
Number of nodes in a cluster	10
Network speed	1Mb/s
Network block size	256 B
Bus speed	25Mb/s
Bus block size	256 B
Local transaction probability	0.712833
Bus-level transaction probability	0.094235
Cluster-level transaction probability	0.092932
X.500 interpretation probability	0.100000
X.500 interpretation rounds	6
CM access probability	0.8

Figure 5.36: Database node utilization, CM probability = 0.8.

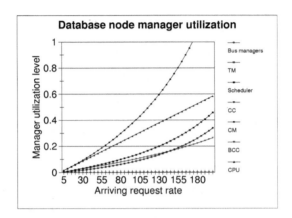

Number of clusters in the system	10
Number of nodes in a cluster	20
Network speed	1Mb/s
Network block size	256 B
Bus speed	25Mb/s
Bus block size	256 B
Local transaction probability	0.712833
Bus-level transaction probability	0.094235
Cluster-level transaction probability	0.092932
X.500 interpretation probability	0.100000
X.500 interpretation rounds	6
CM access probability	0.8

Figure 5.37: Database node utilization, 20 nodes, CM probability = 0.8.

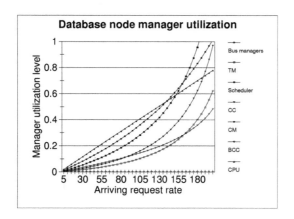

Number of clusters in the system	10
Number of nodes in a cluster	10
Network speed	1Mb/s
Network block size	256 B
Bus speed	25Mb/s
Bus block size	256 B
Local transaction probability	0.712833
Bus-level transaction probability	0.094235
Cluster-level transaction probability	0.092932
X.500 interpretation probability	0.100000
X.500 interpretation rounds	6
CM access probability	0.4

Figure 5.38: Database node utilization, no bus-level commit protocol.

and cluster-level transactions, as was described in Table 5.7. The resulting distribution is that most transactions are local to a single database node. In the analysis shown in Figure 5.39 we used a bus-level transaction probability 0.8.

The result of the changed values is disastrous. Almost all managers overload immediately. Most resources are wasted in management of parallel transactions. It is clear that the Yalin database architecture is not suitable for application environments that need a lot of bus-level transactions.

As a conclusion of the database node analysis, the high proportion of local transactions in the Yalin transaction model allows parallel database nodes to execute at almost full potential. This makes the Yalin cluster architecture quite scalable with IN/GSM classes and transactions. On the other hand, the architecture is not suitable for environments where most cluster-level transactions need bus-level distribution.

Average execution times

After the bottleneck analysis we analyzed average transaction lengths of Location update, Init call, and Query transactions. The set of analyzed transactions covers the properties of Yalin transactions well. We have a short transaction, a regular transaction, and a very long transaction to analyze.

The first analyzed transaction is the Location update transaction. It is the simplest and most common transaction. All it needs to do is to update a mobile equipment location into the Yalin database. The average transaction execution times of the transaction can be seen in Figure 5.40.

The Yalin architecture can handle Location update transactions quite well. Although eventually the transaction execution times will explode, up to 150 service requests may arrive to the system without overload.

The next analyzed transaction is the Init call transaction. It is used whenever a new call is initialized. The analysis can be seen in Figure 5.41. We have three average service times in the transaction: for local, bus, and cluster-level transaction. The Init call transaction may be of any of the three types.

The results of the analysis are not promising in local and cluster level cases. The service

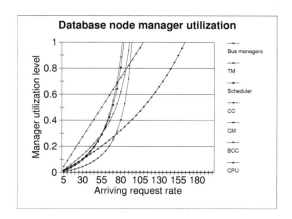

Number of clusters in the system	10
Number of nodes in a cluster	10
Network speed	1Mb/s
Network block size	256 B
Bus speed	25Mb/s
Bus block size	256 B
Local transaction probability	0.157399
Bus-level transaction probability	0.800000
Cluster-level transaction probability	0.020520
X.500 interpretation probability	0.022081
X.500 interpretation rounds	6
CM access probability	0.4

Figure 5.39: Database node utilization, bus-level transaction probability = 0.8,

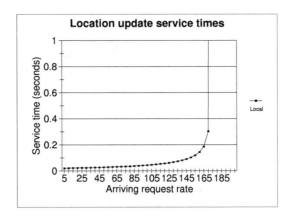

Figure 5.40: Location transaction average service time.

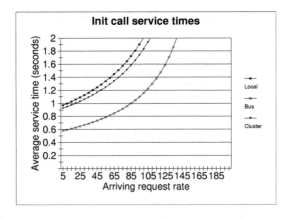

Figure 5.41: Init call transaction average service times.

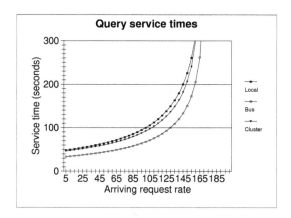

Figure 5.42: Query transaction average service times.

times grow too fast to over two seconds which is too much. The bus-level type behaves better because the transaction calculations may be divided into several database nodes without network level extra costs.

The same trend can be seen in other similar transactions, such as Answer call, charge, various management transactions, and Profile. The service times are around one second each. The problem is not that much in the manager utilizations but more in the number of managers. Each manager needs time to execute its tasks, and combined the times grow this high. However, since this is not an issue of the bottlenecks in the system but rather of the manager service times, the problem can be at least partially corrected by using more efficient CPUs and optimizing disk access.

Finally, the Query transaction is the longest transaction in the Yalin architecture. The results of the Query transaction type analysis are in Figure 5.42. The service times of the Query transactions are around 100 seconds. Also here bus-level parallelism helps in transaction execution times.

While these service times are long, they are not unbearable. The Query transaction is for management alone and even then it is not executed that often. Furthermore, in the analysis we have assumed that Query transactions are long: 1400 CPU operations and 2800 CC operations is assumed to be the average size of a Query type transaction.

Thus, as a result of the time analysis, the transaction execution lengths may be too long for call-related transactions. Other transactions behave well enough. When time is an issue, transaction execution may be optimized for instance by letting a transaction have direct access to nodes without the cluster entry node. Hardware elements may also be optimized. The final way to optimize these transactions is to let them always do reads and writes to main memory data. This definitely helps in service times, but it also makes the architecture more of a main memory database management system than earlier. While main memory is cheap, it is not cheap enough to let all data reside there.

Chapter 6

Conclusion

When new services and service features are introduced and implemented on Intelligent Networks and 3G and 4G wireless networks, also new requirements arise for a database management system that maintains service data. The current database architecture is left open in the ITU-T recommendations. Only the interfaces and efficiency requirements are listed. This thesis helps to clarify database requirements in these specific environments.

The ITU-T recommendations for Intelligent Networks state that the data management system is distributed to several Service data functions, and it must offer real-time access to data. The SDF architecture itself is mostly undefined. It can be implemented on various platforms and still fill the ITU-T recommendations. While distribution is present in the definitions, it is not a necessity in the database implementation.

The data analysis and the Yalin architecture show that IN and wireless network data are compatible at database level. Thus, in the future we can design and implement a single database architecture that supports both platforms. This is a positive result since already 3G and 4G wireless networks wish to use IN for special service creation, use, and management.

In the Yalin architecture analysis we noticed that distribution is actually not useful in the IN/GSM database. When data may be divided into distinct fragments the added computing and disk power helps in transaction throughput. Unfortunately our analysis shows that even a small number of distributed transactions affect database performance on the cluster level. Since X.500 directory information must be updated after data item insertions and deletions, and the update must be reflected to every cluster, we will always have distributed transactions in the system. In a small or a medium telecommunications network, a monolithic database management system may be the best solution for IN/GSM data management.

Where distribution is not useful, parallelism clearly is. Our analysis shows that several parallel database nodes give a better request throughput than a single monolithic node of the same size. Thus, the added CPU and disk capacity is not wasted to node-level atomicity management.

Specialized nodes offer an additional advantage in a parallel database architecture. We can create a Yalin variant where some nodes offer traditional services, some are specialized to mass data services (mass calling and televoting), and some are specialized to VLR services. All transaction requests may arrive via the same cluster entry node. Also the cluster entry node functionality may be expanded to include requests to the X.500 Database entry node which then is a specialized database node in the parallel database architecture.

The next step in Yalin would be a pilot database management system. Yet while the results of the Yalin architecture at analysis level are useful, we are cautious to go into implementation. A database management system of that size would need millions of lines of code. Right now it is not worth the effort. Yet the results of Yalin data analysis, architecture, and bottleneck analysis are useful for any current or future IN or GSM -based database management system design.

The analysis has a value in itself regardless of the IN/GSM results. It clearly shows that our derived queueing model toolbox is useful in transaction-based system analysis. With the toolbox it is possible to create and analyze very complex transaction-based systems and find their bottlenecks. The price for the analysis simplicity is in the analysis details. Since we have used shortcuts whenever possible without losing too much details, the model is best used for the first analysis steps to recognize problematic elements in the system. In the later stages a more detailed analysis with more general (and complex) queueing model tools may be useful.

The queueing model toolbox presented here is only the first step. In the next step we wish to implement a software tool that can be used to create transaction-based models that are based on our derived formulas. With such a tool it is possible to create and analyze complex transaction-based software for bottlenecks.

Next to the bottleneck analysis, also a transaction execution time analysis is important. In a real-time database management system this implies transactions with priorities. The current analysis toolbox cannot be used for this since it assumes that transactions have the same priority.

In the future we wish to generalize our toolbox model to include transaction type priorities. This way we can better analyze real-time transaction execution times. At first we will use fixed priorities and eventually update the toolbox to allow dynamic priorities when possible.

Altogether the work on this thesis gives a good basis for both IN/GSM database management system design and for transaction-based queueing model analysis. As such the work has been beneficial.

Bibliography

[AGP98]   M. Abdallah, R. Guerraoui, and P. Pucheral. One-phase commit: Does it make sense? In *Proceedings of the 1998 International Conference on Parallel And Distributed Systems*, pages 182–192. IEEE, 1998.

[AHC96]   Y. Al-Houmaily and P. Chrysanthis. In search for an efficient real-time atomic commit protocol. In *Proceedings of the IEEE Real-Time Systems Symposium Work In Progress*, pages 51–54. IEEE, 1996.

[Ahn94]   I. Ahn. Database issues in telecommunications network management. In *ACM SIGMOD proceedings*, pages 37–43. ACM, 1994.

[BB95]   A. Bestavros and S. Braoudakis. Value-cognizant speculative concurrency control. In *Proceedings of the 21th International Conference on Very Large Data Bases (VLDB'95)*, pages 122–133. Morgan Kaufmann, 1995.

[BB96]   A. Bestavros and S. Braoudakis. Value-cognizant speculative concurrency control for real-time databases. *Information Systems*, 21(1):75–101, 1996.

[BHG87]   P. Bernstein, V. Hadzilacos, and N. Goodman. *Concurrency Control and Recovery in Database Systems*. Addison-Wesley, 1987.

[BJMC00]   R. Brennan, B. Jennings, C. McArdle, and T. Curran. Evolutionary trends in Intelligent Networks. *IEEE Communications Magazine*, 38(6):86–93, June 2000.

[BN96]   A. Bestavros and S. Nagy. Value-cognizant admission control for rtdb systems. In

A. Bestavros, editor, *Proceedings of the 17th IEEE Real-Time Systems Symposium*, pages 230–239, 1996.

[BPI+01] L. Becchetti, F. Priscoli, T. Inzerilli, P. Mähönen, and L. Munoz. Enhancing IP service provision over heterogeneous wireless networks: A path toward 4G. *IEEE Communications Magazine*, 39(8):74–81, August 2001.

[BS91] K. Bratbergsengen and T. Sæter. The impact of high reliability, high capacity database servers on telecommunication systems and services. In *Proc. of the 6th World Telecommunication Exhibition and Forum, TeleCom 91*, 1991.

[CCI89] CCITT. Recommendations x.500 to x.521 – data communication networks: Directory. In *CCITT Blue Book Volume VIII – Fascicle VIII.8*, pages 1–225. CCITT, 1989.

[CCI92a] *Intelligent Network - Global Functional Plane Architecture - Recommendation I.329 / Q.1203*, 1992.

[CCI92b] CCITT. *Intelligent Network – Service Plane Architecture - Recommendation I.328 / Q.1202*, 1992.

[CDW93] L. B. Cingiser DiPippo and V. Wolfe. Object-based semantic real-time concurrency control. In *Real-Time Systems Symposium Conference Proceedings*, pages 87–96. IEEE, 1993.

[CP84] S. Ceri and G. Pelagatti. *Distributed Databases - Principles and Systems*. McGraw-Hill, 1984.

[DF01] M. Dinis and J. Fernandes. Provision of sufficient transmission capacity for broadband mobile multimedia: A step toward 4G. *IEEE Communications Magazine*, 39(8):46–55, August 2001.

[EGLT76] K. Eswaran, J. Gray, R. Lorie and I. Traiger. The Notions of Consistency and Predicate Locks in a Database System. *Communications of the ACM*, 19(11):624–633, November 1976.

[FGSW00] M. Finkelstein, J. Garrahan, D. Shrader, and G. Weber. The future of the Intelligent Network. *IEEE Communications Magazine*, 38(6):100–107, June 2000.

[GHRS96] R. Gupta, J. Haritsa, K. Ramamritham, and S. Seshadri. Commit processing in distributed real-time database systems. In *IEEE Real-Time Systems Symposium*, pages 220–229. IEEE, 1996.

[GLPT76] J. N. Gray, R. A. Lorie, G. R. Putzolou, and I. L. Traiger. Granularity of locks and degrees of consistency in a shared data base. In G. M. Nijssen, editor, *Modeling in Data Base Management Systems*, pages 365–394, 1976.

[Gra78] J. N. Gray. *Notes on Database Operating Systems*, pages 393–481. Springer-Verlag, 1978.

[GRKK93] J. Garrahan, P. Russo, K. Kitami, and R. Kung. Intelligent Network overview. *IEEE Communications Magazine*, 31(3):30–36, 1993.

[HCL92] J. Haritsa, M. Carey, and M. Livny. Data access scheduling in firm real-time database systems. *Journal of Real-Time Systems*, 4(3):203–241, 1992.

[Hig89] W. Highleyman. *Performance Analysis of Transaction Processing Systems*. Prentice Hall, Englewood Cliffs, New Jersey 07632, 1989.

[HR00] J. Haritsa and K. Ramamritham. Adding PEP to real-time distributed commit processing. In *Proceedings of the 21st Real-Time Systems Symposium*, pages 37–46. IEEE, 2000.

[HRG00] J. Haritsa, K. Ramamritham, and R. Gupta. The PROMPT real-time commit protocol. *IEEE Transactions on Parallel and Distributed Systems*, 11(2):160–181, February 2000.

[IT93a] ITU-T. *Intelligent Network Distributed Functional Plane Architecture - Recommendation Q.1204*, 1993.

[IT93b]   ITU-T. *Intelligent Network Physical Plane Architecture - Recommendation Q.1205*, 1993.

[ITU93a]   ITU-T. *Global Functional Plane for Intelligent Network CS-1 - Recommendation Q.1213*, 1993.

[ITU93b]   ITU-T. *Introduction to Intelligent Network Capability Set 1 - Recommendation Q.1211*, 1993.

[ITU96]   ITU-T. *Distributed Functional Plane for Intelligent Network CS-2 - Recommendation Q.1224 (Draft)*, 1996.

[Jab92]   B. Jabbari. Intelligent Network concepts in mobile communication. *IEEE Communications Magazine*, 30(2):64–69, 1992.

[Kon00]   W. Konhäuser. Mobile communications on the way towards Internet on air. In *Proceedings of WCC-ICCT 2000 Communication Technology, volume 2*, pages 993–1000. IEEE, 2000.

[Lin00]   J. Lindström. Extensions to optimistic concurrency control with time intervals. In D. C. Young, editor, *Proceedings of the 7th International Conference on Real-Time Computing Systems and Applications*, pages 108–115, 2000.

[LNPR00]   J. Lindström, T. Niklander, P. Porkka, and K. Raatikainen. A distributed real-time main-memory database for telecommunication. In *Databases in Telecommunications; International Workshop, Co-located with VLDB-99*, 2000.

[LS93]   J. Lee and S. H. Son. Using dynamic adjustment of serialization order for real-time database systems. In *Real-Time Systems Symposium Conference Proceedings*, pages 66–75. IEEE, 1993.

[LS94]   J. Lee and S. Son. Semantic-based concurrency control for object-oriented database systems supporting real-time applications. In *6th IEEE Euromicro Workshop on Real-Time Systems*, pages 156–161. IEEE, 1994.

[LW95]   P. Lehtinen and A. Warsta. Nokia's IN solution for fixed and cellular networks.
            In Jarmo Harju, Tapani Karttunen, and Olli Martikainen, editors, *Intelligent Networks*, pages 61–67. Chapman & Hall, 1995.

[MC00]   M. Mampaey and A. Couturier. Using TINA concepts for in evolution. *IEEE Communications Magazine*, 38(6):94–99, June 2000.

[Mit95]   H. Mitts. Use of Intelligent Networks in the universal mobile telecommunication
            system (UMTS). In Jarmo Harju, Tapani Karttunen, and Olli Martikainen, editors,
            Intelligent Networks, pages 236–245. Chapman & Hall, 1995.

[MP92]   M. Mouly and M. Pautet. *The GSM System for Mobile Communications*. Mouly &
            Pautet, 1992.

[NB00]   K. Nørvåg and K. Bratbergsengen. Optimal object descriptor caching in temporal
            object database systems. In K. Tanaka, S. Ghandeharizadeh, and Y. Kambayashi,
            editors, *Information Organization and Databases*, pages 235–248, Boston, Mass.,
            2000. Kluwer.

[OV99]   M. Tamero Ozsu and P. Valduriez. *Principles of Distributed Database Systems -
            Second Edition*. Prentice Hall International Inc., 1999.

[Pan95]   R. Pandya. Emerging mobile and personal communication systems. *IEEE Communications Magazine*, 33(6):44–52, 1995.

[PMHG95]   G. Pollini, K. Meier-Hellstern, and D. Goodman. Signaling traffic volume generated by mobile and personal communications. *IEEE Communications Magazine*,
            33(6):60–65, 1995.

[Pon93]   C. Pontailler. TMN and new network architectures. *IEEE Communications Magazine*, 31(4):84–88, 1993.

[Pra99]   N. Prasad. GSM evolution towards third generation UMTS/IMT2000. In *IEEE
            Conference on Personal Wireless Communication*, pages 50–54. IEEE, 1999.

[Raa94]   K. Raatikainen. Information aspects of services and service features in intelligent
            network capability set 1. Report C-1994-45, University of Helsinki, Dept. of Com-
            puter Science, Helsinki, Finland, September 1994.

[Ram93]   K. Ramamritham. Real-time databases. *Distributed and Parallel Databases*,
            1(2):199–226, 1993.

[Ric00]   K. Richardson. UMTS overview. *IEEE Electronics & Communications Engineer-
            ing Journal*, 12(3):93–100, June 2000.

[RKMT95]   K. Raatikainen, T. Karttunen, O. Martikainen, and J. Taina. Evaluation of database
            architectures for intelligent networks. In *Proceedings of the 7th World Telecom-
            munication Forum (Telecom 95), Technology Summit, Volume 2*, pages 549–553,
            Geneva, Switzerland, September 1995. ITU.

[RKV+01]   T. Robles, A. Kadelka, H. Velayos, A. Lappetelainen, A. Kassler, H. Li, D.
            Mandato, J. Ojala, and B. Wegmann. QoS support for an all-IP system beyond
            3G. *IEEE Communications Magazine*, 39(8):64–73, August 2001.

[SC90]   J. Stamos and F. Cristian. A low-cost atomic commit protocol. In *Proceedings
            of the 9th IEEE Symposium on Reliable Distributed Systems*, pages 66–75. IEEE,
            1990.

[Ske81]   D. Skeen. Nonblocking commit protocols. In Y. Edmund Lien, editor, *Proceedings
            of the 1981 ACM SIGMOD International Conference on Management of Data, Ann
            Arbor, Michigan, April 29 - May 1, 1981*, pages 133–142. ACM Press, 1981.

[SRL90]   L. Sha, R. Rajkumar, and J. Lehoczky. Priority inheritance protocols: An approach
            to real-time synchronization. *IEEE Transactions on Computers*, 39(9):1175–1185,
            1990.

[SRSC91]   L. Sha, R. Rajkumar, S. Son, and C. Chang. A real-time locking protocol. *IEEE
            Transactions on Computers*, 40(7):793–800, 1991.

[Tai94a]   J. Taina. Problem classes in intelligent network database design. In O. Martikainen
            and J. Harju, editors, *Proceedings of IFIP TC6 Workshop on Intelligent Networks*,
            pages 194–207, Lappeenranta, Finland, August 1994. Lappeenranta University of
            Technology, Chapman & Hall.

[Tai94b]   J. Taina. Evaluation of OMG, ODMG, X.500, and X.700 data models. Technical
            report, University of Helsinki, Department of Computer Science, 1994.

[TNA+01]   T. Tjelta, A. Nordbotten, M. Annoni, E. Scarrone, S. Bizzarri, Laurissa Tokarchuk,
            John Bigham, Chris Adams, Ken Craig, and Manuel Dinis. Future broadband radio
            access systems for integrated services with flexible resource management. *IEEE
            Communications Magazine*, 39(8):56–63, August 2001.

[TR96]   J. Taina and K. Raatikainen. Experimental real-time object-oriented database ar-
            chitecture for intelligent networks. *Engineering Intelligent Systems*, 4(3):57–63,
            September 1996.

[TR00]   J. Taina and K. Raatikainen. Requirements analysis of distribution in databases
            for telecommunications. In Willem Jonker, Peter Apers, and Tore Sæter, editors,
            Proceedings of the Workshop on Databases in Telecommunications 1999, pages
            74–89. Springer, 2000.

[TRR95]   J. Taina, M. Rautila, and K. Raatikainen. An experimental database architecture
            for intelligent networks. In *Proceedings of IFIP IN'95 Conference*. IFIP, 1995.

[TS96]   J. Taina and S. Son. Requirements for real-time object-oriented database models–
            how much is too much? In *Proceedings of the Ninth Euromicro Workshop on
            Real-Time Systems*, pages 258–265. IEEE, 1996.

[TS97a]   J. Taina and S. Son. TARTOS - toolbox for active real-time object-oriented
            database system models. In *IEEE Workshop on Parallel And Distributed Systems
            (WPDRTS'97)*, pages 131–140. IEEE, 1997.

[TS97b]   J. Taina and S. Son. A framework for real-time object-oriented database models. In *International Workshop on Object-Oriented Real-Time and Dependable Systems (WORDS'97)*, pages 146–152. IEEE, 1997.

[YK00]   J. Yang and I. Kriaras. Migration to all-IP based umts networks. In *Proceedings of the First International Conference on 3G Mobile Communication Technologies*, pages 19–23. IEEE, 2000.

[YWLS94] P. Yu, K. Wu, K. Lin, and S. Son. On real-time databases: Concurrency control and scheduling. *Proceedings of the IEEE*, 82(1):140–157, 1994.

VDM
Verlag
Dr. Müller

Wissenschaftlicher Buchverlag bietet

kostenfreie

Publikation

von

wissenschaftlichen Arbeiten

Diplomarbeiten, Magisterarbeiten, Master und Bachelor Theses
sowie Dissertationen, Habilitationen und wissenschaftliche Monographien

Sie verfügen über eine wissenschaftliche Abschlußarbeit zu aktuellen oder zeitlosen
Fragestellungen, die hohen inhaltlichen und formalen Ansprüchen genügt,
und haben **Interesse an einer honorarvergüteten Publikation**?

Dann senden Sie bitte erste Informationen über Ihre Arbeit per Email
an info@vdm-verlag.de. Unser Außenlektorat meldet sich umgehend bei Ihnen.

VDM Verlag Dr. Müller Aktiengesellschaft & Co. KG
Dudweiler Landstraße 125a
D - 66123 Saarbrücken

www.vdm-verlag.de

www.ingramcontent.com/pod-product-compliance
Lightning Source LLC
LaVergne TN
LVHW022318060326
832902LV00020B/3532